The Palgrave Macmillan Animal Ethics Series

Series editors: **Andrew Linzey** and **Priscilla Cohn**

In recent years, there has been a growing interest in the ethics of our treatment of animals. Philosophers have led the way, and now a range of other scholars have followed from historians to social scientists. From being a marginal issue, animals have become an emerging issue in ethics and in multidisciplinary inquiry. This series explores the challenges that Animal Ethics poses, both conceptually and practically, to traditional understandings of human-animal relations.

Specifically, the Series will:

- provide a range of key introductory and advanced texts that map out ethical positions on animals;
- publish pioneering work written by new, as well as accomplished, scholars, and
- produce texts from a variety of disciplines that are multidisciplinary in character or have multidisciplinary relevance

Titles include

AN INTRODUCTION TO ANIMALS AND POLITICAL THEORY
Alasdair Cochrane

CRITICAL ANIMAL STUDIES: AN INTRODUCTION
Dawne McCance

AN INTRODUCTION TO ANIMALS AND THE LAW
Joan E. Schaffner

Forthcoming titles

ANIMALS, SUFFERING AND PHILOSOPHY
Elisa Aaltola

HUMANS AND ANIMALS: THE NEW PUBLIC HEALTH PARADIGM
Aysha Akhtar

HUMAN ANIMAL RELATIONS: THE OBLIGATION TO CARE
Mark Bernstein

ANIMAL ABUSE AND HUMAN AGGRESSION
Eleonora Gullone

ANIMALS IN THE CLASSICAL WORLD: ETHICAL PERCEPTIONS
Alastair Harden

ANIMAL SUFFERING AND THE PROBLEM OF EVIL
Nicola Hoggard Creegan

POWER, KNOWLEDGE, ANIMALS
Lisa Johnson

ANIMAL EXPERIMENTS: EVIDENCE AND ETHICS
Andrew Knight

POPULAR MEDIA AND ANIMAL ETHICS
Claire Molloy

ANIMALS, EQUALITY AND DEMOCRACY
Siobhan O'Sullivan

SOCIAL WORK AND ANIMALS: A MORAL INTRODUCTION
Thomas Ryan

Also by Joan E. Schaffner

LITIGATING ANIMAL LAW DISPUTES (editor with Julie Fershtman, 2009)

A LAWYER'S GUIDE TO DANGEROUS DOG ISSUES (editor, 2009)

The Palgrave Macmillan Animal Ethics Series
Series Standing Order ISBN   **978–0–230–57686–5 Hardback**
                                978–0–230–57687–2 Paperback
(outside North America only)

You can receive future titles in this series as they are published by placing a standing order.
Please contact your bookseller or, in case of difficulty, write to us at the address below with
your name and address, the title of the series and the ISBN quoted above.

Customer Services Department, Macmillan Distribution Ltd, Houndmills, Basingstoke,
Hampshire RG21 6XS, England

An Introduction to Animals and the Law

Joan E. Schaffner

Associate Professor, George Washington University Law School

First published 2011 by
PALGRAVE MACMILLAN

Palgrave Macmillan in the UK is an imprint of Macmillan Publishers Limited,
registered in England, company number 785998, of Houndmills, Basingstoke,
Hampshire RG21 6XS.

Palgrave Macmillan in the US is a division of St Martin's Press LLC,
175 Fifth Avenue, New York, NY 10010.

Palgrave Macmillan is the global academic imprint of the above companies
and has companies and representatives throughout the world.

Palgrave® and Macmillan® are registered trademarks in the United States,
the United Kingdom, Europe and other countries

ISBN   978–0–230–23563–2   hardback
ISBN   978–0–230–23564–9   paperback

This book is printed on paper suitable for recycling and made from fully
managed and sustained forest sources. Logging, pulping and manufacturing
processes are expected to conform to the environmental regulations of the
country of origin.

A catalogue record for this book is available from the British Library.

A catalog record for this book is available from the Library of Congress.

*To my mother, in memory of my father, and
with love and affection for my furry and feathered family.*

Contents

Abbreviations

Case Reporters

A. Atlantic Reporter, State Appellate Court Decisions for Connecticut, Delaware, District of Columbia, Maine, Maryland, New Hampshire, New Jersey, Pennsylvania, Rhode Island, Vermont
A.D. New York Appellate Division Reports
Eur. Ct. H.R European Court of Human Rights
F. Federal Reporter, United States Courts of Appeals
F. Supp Federal Supplement, United States District Courts
H.C.J. Israeli High Court of Justice
Il. App. Illinois Appellate Court Reports
Mass. Massachusetts Reports, Massachusetts Supreme Court
Misc. New York Miscellaneous Reports
N.E. North Eastern Reporter, State Appellate Court Decisions for Illinois, Indiana, Massachusetts, New York, Ohio
N.J. New Jersey Reports, N.J. Supreme Court
NSWSC New South Wales Supreme Court (Australia)
N.W. North Western Reporter, State Appellate Court Decisions for Iowa, Michigan, Minnesota, Nebraska, North Dakota, South Dakota, Wisconsin
N.Y.S. New York Supplement, New York Supreme Court, Appellate Division
P. Pacific Reporter, State Appellate Court Decisions for Alaska, Arizona, California, Colorado, Hawaii, Idaho, Kansas, Montana, Nevada, New Mexico, Oklahoma, Oregon, Utah, Washington, Wyoming
S. Ct. Supreme Court Reporter, United States Supreme Court
So. Southern Reporter, State Appellate Court Decisions for Alabama, Florida, Louisiana, Mississippi
S.W. Southwestern Reporter, State Appellate Court Decisions for Arkansas, Kentucky, Missouri, Tennessee, Texas
U.S. United States Reports, United States Supreme Court
W.L. Westlaw

Regulations, Statutes and Treaties

C.F.R. United States Code of Federal Regulations
Conn. Gen. Stat. General Statutes of Connecticut
D.C. Stat. District of Columbia Statutes (Code)
Fl. Stat. Florida Statutes
I.L.C.S. Illinois Compiled Statues
Md. Code Maryland Code
Minn. Stat. Minnesota Statutes
N.C. Gen. Stat. General Statutes for North Carolina
N.M. Stat. New Mexico Statutes
R.C.W. Revised Code of Washington State
Tenn. Code Tennessee Code
U.S.C. United States Code
Va. Code Ann. Virginia Code Annotated

Table of Authorities

Constitutions

Cases

Regulations, Statutes and Treaties

United States

Federal

United Kingdom

Australasia

Series Preface

This is a new book series for a new field of inquiry: Animal Ethics.

In recent years, there has been a growing interest in the ethics of our treatment of animals. Philosophers have led the way, and now a range of other scholars have followed from historians to social scientists. From being a marginal issue, animals have become an emerging issue in ethics and in multidisciplinary inquiry.

In addition, a rethink of the status of animals has been fuelled by a range of scientific investigations which have revealed the complexity of animal sentiency, cognition and awareness. The ethical implications of this new knowledge have yet to be properly evaluated, but it is becoming clear that the old view that animals are mere things, tools, machines or commodities cannot be sustained ethically.

But it is not only philosophy and science that are putting animals on the agenda. Increasingly, in Europe and the United States, animals are becoming a political issue as political parties vie for the "green" and "animal" vote. In turn, political scientists are beginning to look again at the history of political thought in relation to animals, and historians are beginning to revisit the political history of animal protection.

As animals grow as an issue of importance, so there have been more collaborative academic ventures leading to conference volumes, special journal issues, indeed new academic animal journals as well. Moreover, we have witnessed the growth of academic courses, as well as university posts, in Animal Ethics, Animal Welfare, Animal Rights, Animal Law, Animals and Philosophy, Human-Animal Studies, Critical Animal Studies, Animals and Society, Animals in Literature, Animals and Religion – tangible signs that a new academic discipline is emerging.

"Animal Ethics" is the new term for the academic exploration of the moral status of the non-human – an exploration that explicitly involves a focus on what we owe animals morally, and which also helps us to understand the influences – social, legal, cultural, religious and political – that legitimate animal abuse. This series explores the challenges that Animal Ethics poses, both conceptually and practically, to traditional understandings of human-animal relations.

The series is needed for three reasons: (i) to provide the texts that will service the new university courses on animals; (ii) to support the increasing number of students studying and academics researching in animal related fields, and (iii) because there is currently no book series that is a focus for multidisciplinary research in the field.

Specifically, the series will

- provide a range of key introductory and advanced texts that map out ethical positions on animals;
- publish pioneering work written by new, as well as accomplished, scholars, and
- produce texts from a variety of disciplines that are multidisciplinary in character or have multidisciplinary relevance.

The new Palgrave Macmillan Series on Animal Ethics is the result of a unique partnership between Palgrave Macmillan and the Ferrater Mora Oxford Centre for Animal Ethics. The series is an integral part of the mission of the Centre to put animals on the intellectual agenda by facilitating academic research and publication. The series is also a natural complement to one of the Centre's other major projects, the *Journal of Animal Ethics*. The Centre is an independent "think tank" for the advancement of progressive thought about animals, and is the first Centre of its kind in the world. It aims to demonstrate rigorous intellectual enquiry and the highest standards of scholarship. It strives to be a world-class centre of academic excellence in its field.

We invite academics to visit the Centre's website www. oxfordanimalethics.com and to contact us with new book proposals for the series.

Andrew Linzey and Priscilla N. Cohn
General Editors

Acknowledgments

This text was first conceived when I heard of the launch of the Palgrave Macmillan Series on Animal Ethics in partnership with the Ferrater Mora Oxford Centre for Animal Ethics. As a fellow with the Centre, I proposed this text to Andrew Linzey and Priscilla Cohn, who graciously accepted my proposal. I am so grateful for this extraordinary opportunity to author, as part of the series, the introductory text addressing the law and its treatment of nonhuman animals. It is with great enthusiasm and humility that I add my voice to the experts who have preceded me and from whom I have learned so much.

The area of animal law as a distinct practice in the United States was conceived in the 1970s by remarkable pioneers including David Favre, Gary Francione, Henry Mark Holzer, Larry Kessenik, Sara Luick, Joyce Tischler, and Steven Wise. In fact, the Animal Legal Defense Fund, founded by these pioneers and others, celebrated its 30th anniversary in August 2009. Over the past three decades many outstanding lawyers have followed in their footsteps, expanding the field and promoting the interests of animals through the law. The animal law movement brings together legal academics, practitioners, and grassroots volunteers, working together to promote the interests of animals through the law. As a relative newcomer to this area I am grateful to these individuals and others for their insight, tenacity, and passion. This text draws from their work to present an overview of animal law, to identify the strengths and weaknesses of the current legal regime, and to explore ways in which the law can better protect the interests of animals.

I must disclose the personal bias I bring to this endeavor. My deep love and appreciation for all animals informs my views of the laws designed to protect them. A theme that runs throughout this text is that the law, for the most part, is human-centered, designed primarily to protect human interests. This is not a novel realization and should not strike anyone as unusual since the law is drafted by humans and for humans. The first generation of lawyers and scholars struggled to alter the law to recognize that nonhuman animals are sentient beings and should be, at a minimum, protected from cruelty and neglect. Today even that fight has not been won for all animals. However, societal views of animals and our relationships to them continue to change. It is time for the next generation of lawyers not only to continue to fight for minimal protections of all animals, independent of their relationship with humans, but to develop an all-inclusive legal regime that includes animals as subjects and appreciates, respects, and protects animals' inherent interests in their own right.

I thank all who have worked tirelessly to promote legal protections for animals and paved the way for the second generation. I also want to thank my colleagues who have provided invaluable insight and support to me in drafting this text, including Wendy Anderson, Priscilla Cohn, Claudia Haupt, Andrew Linzey, and Nathan Winograd. Finally, I offer special thanks to my outstanding research assistant, Laura Verlangieri, and the students in my Animal Law Lawyering seminar who waded through this text at an early stage and whose comments and papers provided much insight. They are Halli Bayer, Alison Bruenjes, Andrew Friedman, Julie Jankowski, Sara Marshman, and Debodhonyaa Sengupta.

1
Animals and the Law: The Basics

Introduction

Animals and their treatment by humans appear in our law in a variety of ways. Consider Matthew, a 27-year-old chimp whose case was accepted by the European Court of Human Rights in 2008.[1] Animal rights activist and teacher Paula Stibbe wanted to be declared his legal guardian should the sanctuary where he resides close due to bankruptcy. Austria's Supreme Court upheld the lower court ruling rejecting her request. The problem, under Austrian law, is that only legal "persons" may be appointed a guardian. Since Matthew is deemed a legal "thing" under Austrian law he cannot be appointed a guardian. In contrast, in 2007 when Michael Vick, an American football star, was arrested for dog-fighting and his "pit bull type" dogs were seized in the civil forfeiture action associated with the criminal case, the government appointed a Special Master/Guardian for the dogs.[2] Although animals are deemed "things" under state and federal law in the United States, the court held that such status would not prohibit assigning a human to represent the dogs' interests.[3]

Another example is animals that we raise and slaughter for food. The 1965 Brambell Report,[4] a United Kingdom government publication, addressed the general concept of the welfare of animals used in farming in the United Kingdom and the likely affect of intensive husbandry practices on the animals' welfare. Such practices, called factory farms, are defined by Webster's Dictionary as "a system of large-scale industrialized and intensive agriculture that is focused on profit with animals kept indoors and restricted in mobility." This ground-breaking report used the concepts of sentience and animal welfare for animals used in farming. The report, considered the most influential investigation of animal welfare for animals kept intensively in the twentieth century, had a beneficial impact on the welfare of many millions of animals, and resulted in the establishment of regulations and the *Codes of Recommendations for the*

Welfare of Livestock in the United Kingdom.[5] In contrast, animals used in farming in the United States have no protection against inhumane treatment on the farm. Moreover, less than 10 percent of these animals receive only the most minimal of protections during transport and slaughter because turkeys, chickens, and other fowl ("poultry") who account for over 90 percent of all animals used for farming are exempt from the humane transport and slaughter laws and thus receive no protection at all.

Yet another example comes from Spain where, in 2008, the Deputies of Environment, Agriculture and Fishing Commission of the Spanish Parliament approved support of the Great Ape Project (GAP)[6] with the expectation that it would be implemented into law within four months.[7] The law would grant chimpanzees, gorillas, orangutans, and bonobos, the right to life, protection of individual liberty, and protection from torture.[8] The law also would prevent the use of great apes for research or entertainment but allow zoos currently housing great apes to retain them so long as conditions are improved.[9] In contrast, bull-fighting is not only legal but a popular pastime for some citizens in Spain. Each year, bull-fighting season is literally kicked-off with the running of the bulls in Pamplona being taunted by hundreds of humans. Every weekend bulls are slaughtered for mere sport or what some call art. What explains this dichotomy? Some may argue that the proven highly developed intelligence and emotional capacity of the great apes, the closest species relative to humans, supports rights for them and not for bulls. Others may justify different treatment based on the fact that bull-fighting turns a profit for some and is deeply ingrained in Spanish culture and tradition. However, bulls and apes are sentient creatures and arguably both deserve to be free from pain. In fact the Parliament of Catalonia in Barcelona, an autonomous region of Spain, arguably agreed and is considering a ban on bull-fighting.[10] The owner of Barcelona's last major bullring described the ban as "an attack on the freedom" of those who love bullfights.[11] While the central government of Spain in Madrid does not favor a ban, it will respect the Catalan decision.

In contrast to these nations, China has virtually no law protecting animals. Examples of the mistreatment of animals in China include "live dogs and cats crammed into tiny cages for sale as meat in markets, live cattle and goats being fed to zoo animals for public entertainment, and dogs being skinned alive for fur production."[12] In 2004, the Beijing government announced the draft of animal welfare legislation.[13] Soon thereafter it was withdrawn from consideration. People believed the law was impractical and premature. However, a very brief chapter on animal welfare was added to the Chinese laws governing "experimental animals." In June 2009, the Hanzhong government brutally killed over 36,000 dogs after a rabies outbreak.[14] In response to this brutal killing, the central government established an animal protection legislation project panel of legal experts to draft a comprehensive animal welfare law in consultation with the Royal Society for the Prevention of Cruelty to Animals (RSPCA). The draft was unveiled in September 2009 and included provisions governing the welfare of companion animals and animals used for food, research and entertainment. However, by January 2010, the law had been criticized by many arguing that given the current state of affairs for humans in China, human welfare should be the first priority. In response, the panel plans to propose laws that will concentrate on the prohibition of animal cruelty, including a ban on the use of cats and dogs for food.[15]

This introductory text is designed to explore this diverse, inconsistent, and controversial area of law, animal law. The text will present a basic survey of the laws that are supposed to protect animals' interests, and analyze and critique them. As the above examples demonstrate, humans' legal relationship to animals is complex and varied. Currently, the law criminalizes deliberate individual acts of gratuitous cruelty towards most animals, yet allows and even supports institutionalized cruelty of animals. Because the law is drafted by humans and for humans it virtually always favors human interests over the interests of all other species. *Arbitrarily* favoring one species' interests over the interests of another is speciesist.[16] The term "speciesism" was first used in 1970 by Richard Ryder, a psychologist, and since then animal activists and others have used the term to describe various forms of discrimination against sentient nonhumans. Peter Singer defined speciesism in *Animal Liberation* as "the idea that it is justifiable to give preference to beings simply on the grounds that they are members of the species *Homo sapiens.*"[17] However, this definition "limits speciesism to bias in favor of only one species."[18] Rather, speciesism more generally is a failure to give equal consideration to any sentient being purely on the basis of the species to which that sentient being belongs. Note that virtually all theoreticians, advocates, and others agree that *sentience*, generally understood as the ability to feel or perceive pleasure or pain, is a relevant criterion on which to distinguish species when determining which beings deserve moral consideration. However, beyond this minimum threshold, many still debate whether any other characteristic is necessary to warrant moral consideration of a being. The premise of this text is that sentience is the only necessary criterion for awarding moral consideration to a being. Thus, for the purposes of this text, references in general to "beings," "animals" and "species," unless otherwise noted, refer to those who are sentient.

Our current law is not only speciesist but anthropocentric. It addresses reality primarily if not exclusively in terms of human values and experiences. This human-centered approach appears based on a fundamental belief that humans are the *only* species worthy of respect as subjects under the law. Furthermore, the law is anthropomorphic, failing to appreciate the inherently unique and different qualities of all other species and instead attributing human motivation, characteristics, or behavior to non-human animals. Andrew Linzey, an animal theologian, in his book *Why Animal Suffering Matters*,[19] claims there is good and bad anthropomorphism. Bad anthropomorphism primarily understands animals in terms of our own human needs and emotions. Good anthropomorphism accepts that all sentient beings suffer to a greater or lesser extent than we do.[20] As an example of good anthropomorphism Linzey cites the 1970 minority position of the former United Kingdom Farm Animal Welfare Advisory Committee stating: "the fact that an animal has limbs should give it the right to use them; the fact that a bird has wings should give it the right to spread them; the fact that both animals and birds are mobile should give them the right to turn around, and the fact that they have eyes should give them the right to see."[21]

Furthermore, we do not need science to prove what can be reasonably assumed given our knowledge of our own suffering. Given what we know about sentience it is only reasonable to suppose that sentient beings would be harmed in situations that would harm us or would be hampered when their natural abilities are frustrated, like us. Thus, if the law is to be non-speciesist it must not depend solely on species membership, but must give equal consideration to comparable interests of all sentient species. Moreover, the law must appreciate and value the inherently unique qualities of all sentient species to avoid being anthropocentric and anthropomorphic. A focus of this text is to identify the speciesist, anthropocentric, and anthropomorphic aspects of the current law and discuss ways to avoid these tendencies in order to properly protect the interests of all sentient beings under the law.

What is animal law?

Throughout history humans have raised and confined animals for food, clothing and research, trained animals for entertainment, fought animals for sport, bought and sold animals for profit, and lived with animals for companionship. Animal law, in the broadest sense, is legal doctrine in which the legal, social or biological nature of nonhuman animals is an important factor.[22] This area of law is quite diverse and cuts across every substantive area of the law including property, tort, contract, criminal, family, and trusts; all jurisdictional boundaries – federal, state and international; and every source of law – constitutional, statutory, regulatory and common law.

Lawyers who are interested in improving the legal status and/or treatment of animals such that their interests and inherent worth are recognized and protected directly under the law focus on a subset of animal law which I will call "animal protection" law. The primary difference between these lawyers and other lawyers who handle cases involving animals is that the focus of these lawyers (and this text) is on the animal's own interests rather than the human owner's interests in their animal. Although this text focuses primarily on the laws of the United States,[23] the laws of other nations, including the United Kingdom,[24] Australasia,[25] and the European Union,[26] are included for comparison and contrast and to provide a somewhat global introduction to animal protection law. Since this text is introductory, it is not a comprehensive treatise on the animal laws of each nation. Rather, the goal is to survey the primary laws that protect animals, identify the themes that link them, analyze and critique them in light of their consideration and protection of animals' interests, and explore characteristics of a future legal system that would better protect animals' interests.

As we will see, the terminology used to describe animals under the law is important both from a policy standpoint as well as a legal standpoint. Language is not neutral but rather shapes our perception of the world. Throughout this text, we will note current terms used in the law to describe animals and explain how these terms are often derogatory and shape our views of animals. Terms such as "brutes," "dumb creatures" and "pests," for example, denigrate animals and prevent the law from adequately accounting for their inherent value. Moreover, the law references animals based on their use by humans. As an example, animals who are used in farming are referred to as "farm animals" or those who are used for research as "research animals." This terminology implies that their only value is their worth to humans as objects of food or research. This text, when not quoting sources, will use ethically sensitive language when describing animals and their relationship to humans. The hope is to instill a greater appreciation for animals' inherent value by referring to them in a manner that reflects that appreciation and reshapes our perception of them.

This introductory chapter presents a very basic introduction to "animals" and "law." Although many will find they are quite familiar with these basics, the information is fundamental to the materials discussed in later chapters. Later chapters will discuss how the law, if at all, protects animals. These chapters are organized around the nature of the law that governs – the criminal anti-cruelty laws, the statutory and regulatory animal welfare laws and animal management and control laws, the constitution, and the private statutory and common law. The final chapter will explore possibilities for the future of animal protection law.

The "animals"

Animals are beings that comprise a multitude of species each with very different sets of capabilities, including flight, breathing, senses of smell, sight and hearing, and means of communicating. Of course, humans are animals. However, when "animals" are referenced, whether it be in law or in general parlance, the reference, generally, is to every species except humans. Nevertheless, the fact that humans are animals is particularly important to an inquiry into animal law, as humans comprise a subset of the group under study. In fact, Charles Darwin, over 150 years ago, "maintained that there are no uniquely human characteristics. The difference in mind between man and the higher animals, great as it is, is certainly one of degree and not of kind."[27] Thus, according to Darwin, the difference between human animals and nonhuman sentient animals is quantitative, differing as to the level of a given attribute, not qualitative, differing in the kind of attributes we have. Nevertheless, for purposes of this text, the term "animals" will refer to *nonhuman* sentient animals unless otherwise noted.

Cornell University historian Dominick LaCapra claims the twenty-first century will be the century of the animal.[28] Disciplines ranging from evolutionary biology and cognitive ethology to psychology, anthropology, philosophy, history and religious studies include research into animal intelligence and emotion. Marc Bekoff and Jessica Pierce, in their book *Wild Justice: The Moral Lives of Animals*, assert that

> animals not only have a sense of justice, but also a sense of empathy, forgiveness, trust, reciprocity, and much more ... [The authors maintain] that animals have rich inner worlds – they have a nuanced repertoire of emotions, a high degree of intelligence ... and demonstrate behavioral flexibility as they negotiate complex and changing social relationships.[29]

The authors recount examples of a variety of animal species demonstrating these characteristics. One account is of eleven elephants who rescued an antelope captured by humans in KwaZulu-Natal. The authors explain how the elephants waited until night-time when the humans were asleep and the matriarch then unlatched the gate of the enclosure with her trunk and allowed the antelope to escape. Another example is of the rat in a cage who refused to push a lever for his food when he saw *another* rat being electrically shocked as a result. A third account is of Libby the cat who led her elderly, deaf and blind dog friend, Cashew, away from obstacles to find his food.

Any human who lives with an animal believes that their companion feels joy and sorrow, develops relationships, communicates, and is intuitive and smart. The divide between human and nonhuman animals is indeed much narrower than once believed. Of course, different species of animals have different levels of cognitive, emotional and moral capabilities, but what is clear is that homo sapiens are not the only species with these and other capabilities. On the other hand, nonhuman animals are quite different from human animals. They have many qualities that humans do not possess; for example, the ability to fly or to breathe under water, or heightened senses of smell, sight, and sound, and the apparent ability to predict impending disasters.[30]

What is the relevance of these similarities and differences for purposes of moral consideration and, in turn, the law? Regardless of differences, sentient beings deserve moral consideration and at least some legal rights, whether they are human or nonhuman. Of course, our legal system is created by humans and primarily devoted to protecting human interests. Thus, as a practical and political matter when advocating for legal protections of nonhuman sentient animals, noting the human-like characteristics of animals is persuasive and helps facilitate the award of legal protections. This is because most humans agree that to the extent animals share human qualities and interests they deserve to be given equal consideration of those same interests. Doubtlessly, this argument capitalizes upon the current anthropocentric view of our society and our law. Nevertheless, until this view is altered, an advocate may use it to his or her advantage to achieve the desired goal, a law that gives equal consideration to at least those interests that nonhuman animals share with humans. However, it is important to note that to the extent sentient beings have qualities and interests different from humans they deserve proper consideration and protection of those interests as well.

Controversy exists over whether sentience is enough to qualify a being for moral consideration. For centuries philosophers have argued that because animals are non-rational beings and/or linguistically deficient they do not deserve moral consideration. However, Linzey argues that to the extent these differences are morally relevant they justify *granting* rather than *denying* to animals moral consideration. For example, Saint Thomas Aquinas reasoned that because animals are not rational they suffer less than humans because humans have the capacity to anticipate harm or death which in turn heightens their suffering. However, the opposite also may be true. Consider, says Linzey, the plight of captive animals or animals used for research or food who are often denied the freedom and opportunity to engage in natural behaviors and subjected to a variety of experiences that cause pain and suffering. Because animals are non-rational,

> [t]hey experience the raw terror of not knowing. And since the implication of the argument is that animals live closer to their bodily senses than we do, the frustration of their natural freedoms may well induce more suffering than we allow ... [Moreover,] they have no means of rationalising their deprivation, boredom, and frustration. They have no intellectual means of escaping their circumstances, for example (as far as we can tell), by use of the imagination.[31]

The moral relevance of this difference suggests that animals and humans suffer in a variety of ways and animals' suffering may be greater than humans under similar circumstances. Thus, there may be grounds for supposing that, in comparison to humans, animals suffer more in some situations and less in others. What is not justifiable, however, is supposing that human suffering *always* counts more than the suffering of nonhuman sentient animals.

Similarly, philosophers, such as Michael Leahy, have argued that animals are linguistically deficient and their absence of language "indicates the absence of self-awareness to the extent that creatures without language cannot have any independent interests ... [As a result] *any* human interest regarded as serious by humans, even if it involves gratuitous cruelty [to animals] ... is justified."[32] Linzey takes issue with this argument on two grounds. First, although animals may not use human language, they have their own means of communication that we cannot appreciate:

> We do not know what it is like precisely to howl, or meow, or to create birdsong. We do not know what it is like to fly through the air, or possess extraordinary capacities for sight and sound, as well as a range of other senses vastly superior to our own. It seems a fair assumption

that there are multiple riches – even spiritual ones – inherent in these experiences that are inaccessible to us. . . . [However,] it doesn't follow that their suffering isn't as significant for them *in their own terms* as ours is for us.[33]

Furthermore, and perhaps more importantly, because we cannot communicate with them, they are incapable of consent:

> In law, the concept of consent is ... central. Consent can make the difference between consensual sex and rape or between acts of sadomasochism and torture. Both ethically and legally, the notion of informed consent has become indispensable in assessing the acceptability of a given act ... [Thus,] the absence of the capacity to give consent, informed or otherwise, must logically tell *against* the abuse of animals ... The inability of animals to consent cannot imply a diminished moral obligation on our part. On the contrary, our responsibility increases as we recognise that relevant factor is absent.[34]

Thus, Linzey's argument is that the putative differences ironically reveal a set of characteristics, such as the inability to consent or a lack of comprehension, which provide grounds for extending (rather than depriving) sentient animals moral consideration.

How does the law define "animal"? While it depends upon the context and use, the broadest definition is found in the laws prohibiting cruelty as they are the most basic of the so-called animal protection statutes that supposedly protect animals from individualized intentional and gratuitous pain and suffering. A survey of all 50 states, the District of Columbia, and Canadian anti-cruelty laws[35] demonstrates that there is no one common legal definition of "animal." The following is just a representative sample. The broadest definition is "Every living creature." Most definitions exclude human beings either expressly or by defining animals as "Every living dumb creature" or "all brute creatures and birds." These very broad definitions include all living beings, whether sentient or not. Other definitions are somewhat narrower. These definitions (although not quoted, all expressly exclude humans) include only sentient beings in the definition of animal; for example, "All living and sentient creatures;" "Every living vertebrate;" "Every warm-blooded living creature;" "Mammal, bird, reptile, amphibian, fish, cetacean, and any other superior phyla animal;" "Mammals, birds, reptiles and amphibians;" "Mammal, bird, fish, reptile or invertebrate;" or "Nonhuman living being with a developed nervous system." Several states expressly exclude certain species from the term "animal," although some are sentient, such as "fish, crustacea or molluska;" "fish or pests;" "insect or reptile;" "fowl;" or "reptile or amphibian."

In sum, most anti-cruelty laws include only sentient animals. This appears appropriate as cruelty, the intentional infliction of pain on another being, presupposes that the victim is capable of feeling and perceiving pain. Anti-cruelty laws that limit the definition of animal further to include only animals that share additional characteristics in common with humans arguably reflect the anthropocentric nature of our law.

Note that these definitions are found within the anti-cruelty statutes and thus apply to these laws, only. The definition of "animal" may vary under other laws governing different contexts, for example, the animal welfare or control laws. Thus, it is critical to note the definition that governs the specific legal context referenced.

As a general matter, animals are characterized as "things" under the law in most countries. Notably, the European Union in the Treaty of Amsterdam and most recently in the Lisbon Treaty defined animals as "sentient beings."[36] However, other countries, including the United States, have not legally recognized animals as sentient beings. In these countries, animals may be owned and used by humans solely for human purposes. This fact is critical to an understanding of the legal treatment of animals in every respect. This characterization has very early roots. For example, Aristotle stated in *Politics* that

the other animals exist for the sake of man, the tame for use and food, the wild, if not all, at least the greater part of them, for food, and for the provision of clothing and various instruments. Now if nature makes nothing incomplete and nothing in vain, the inference must be that she has made all animals for the sake of man.[37]

Centuries later, Immanuel Kant held that people have indirect duties to animals, namely that humans should not harm animals, "for he who is cruel to animals becomes hard also in his dealings with men."[38]

The law distinguishes animals primarily along two lines. The first distinction is whether the animal is domestic or wild; for example, free-ranging.[39] However, there is no bright-line separating the two categories. For example, feral cats present unique problems for humans and the law as they are domesticated animals that have gone "wild" in the sense that they are no longer socialized to humans. Moreover, wild animals may be owned by humans and kept in a domestic, captive, setting. The second distinction is based on humans' use of the animal. The laws are organized around the various uses; for example, animals used for companionship, research, entertainment, and farming. Each of these human uses raises different legal issues and thus different legal regimes have developed to govern each. Of course, the same species of animal can fall within a variety of human uses and thus legal constructs. For example, a dog may be a companion animal, may be used for research, or may provide entertainment to humans; a rabbit may be a companion animal, may be raised and killed for food, may be used for research, or exhibition, or may be wild; while an elephant may live in the wild as part of our environment, or may be captive in a circus or zoo to provide entertainment. The law treats the same animal differently depending upon the human's use of the animal.

The "law"

Legal systems

A legal system is a complex set of principles and rules that govern human behavior. The law defines the rights and obligations of the government to its citizens – public law, and among citizens – private law. At the most fundamental level, human laws exist to minimize, eliminate, and at times shift the burden of harm that humans cause others. The rule of law exists as an alternative to the rule of nature based on violence and the survival of the fittest, to promote a peaceful and harm-free society.

The law derives from various sources of authority – international, constitutional, statutory, administrative, and common law – each involving the protection or regulation of different types of interests, rights, and obligations. The characteristics that distinguish the various sources of law include: (1) the drafting entity and its function – a legislature that drafts laws to govern society, a regulatory agency that enacts regulations to implement legislative prerogatives, or a judge who hears cases and decides individual disputes, (2) the scope of authority – governing a nation, a state, or individual parties, (3) the generality of the provisions – global principles as defined in constitutions or detailed specifications as found in regulations, and (4) the manner of enforcement – by a public entity or private individual. Laws governing animals are found in every legal source. The following briefly describes the attributes and purposes of each source of law in order to prepare the reader to understand, analyze, and critique the current legal landscape as it relates to and protects animals' interests. Consider which sources of law – public or private, constitutional, statutory or common law – provide the greatest opportunity for protecting animals and why. Does the current animal law regime properly capitalize on the opportunity presented by each source of law?

Sources of law

International law

International law regulates the behavior of sovereign states not individuals. Especially in this age of globalization, the actions of one nation may have a very real and dramatic impact on other nations. Thus, there must be certain norms that govern the relationship among nations. Three principal methods exist for creating legally binding rules that govern nations: *jus cogens,* customary rules, and treaties.[40] In the late 1960s a category of general international rules emerged that defined certain fundamental values known as peremptory rules or *jus cogens. Jus cogens* both confer community rights and impose community obligations "essential for the protection of fundamental interests of the international community."[41] They protect fundamental human[42] rights by prohibiting war crimes, crimes against humanity, slavery, racial discrimination, genocide, piracy, apartheid and torture. *Jus cogens* are superior to customary rules and treaties. Customary rules "result ... from the convergence of will of all States," a form of tacit agreement, whereas treaties are only applicable to the contracting parties. Customary rules and treaties are of equal rank and status.

The most frequent means of creating international law is through treaties or conventions, the term often used to refer to formal multilateral treaties. Treaties or conventions are agreements reached by sovereign nations that govern their rights and obligations to each other. Most relevant to the field of animal law are the various treaties or conventions concerning wildlife and the environment. In 1993, the term "environment" was defined in the Council of Europe Convention on Civil Liability for Damage Resulting from Activities Dangerous to the Environment as including "natural resources both abiotic and biotic, such as air, water, soil, fauna and flora and the interaction between the same factors; property which forms part of the cultural heritage; and the characteristic aspects of the landscape."[43] In the United States, treaties duly ratified by the President after the Senate's approval are equal in rank to other federal law. However, for treaties containing non-self-executing provisions, implementing legislation must be passed to apply these provisions. "By contrast, [in the United Kingdom] ... treaties do not bind national authorities unless they are translated into detailed national legislation."[44] A treaty drafted in violation of *jus cogens* is null and void.

The United Kingdom is a member state of the European Union (EU) and a member of the Council of Europe. The Council of Europe has a membership of 47 countries and is an intergovernmental organization with the purpose of protecting human rights and promoting democracy and the rule of law throughout Europe. The treaties of the Council are not directly binding on member countries unless ratified by the member country. In contrast, the EU is a formal economic and political partnership among 27 member states. Laws of the EU may be directly binding on member states.[45]

Constitutional law

Constitutional law is the supreme law of a sovereign. It describes the political and legal structure of the government, and sets forth the most fundamental principles of a given society. The United Kingdom and New Zealand have unwritten constitutions, meaning no single document lays out the basic principles of these nations; rather, basic principles are derived from other sources. Many countries, including the United States and Australia, have written constitutions. The Constitution of the United States, for example, creates the federal government and defines its limited powers among three branches – the legislative, executive and judicial branches. Each branch has authority to create law. Statutory law is enacted within the legislative branch and is the primary source of law that regulates the behavior of individuals as well as both private and public entities. Administrative law is created within the executive branch, pursuant to authority granted by the legislature. These regulations implement the will of the legislative branch. Finally, common law is the law developed through judicial decision-making. The power of the federal government is limited, meaning it only has the powers enumerated in the Constitution; all other powers are left to the states. Federal laws governing animals are authorized either by the treaty power of the federal government, art. II, sec. 2, or by the commerce clause that grants federal authority to regulate interstate commerce, art. I, sec. 8, cl. 2.

The Constitution, the Bill of Rights, and later amendments, define the fundamental rights of liberty and justice for all human citizens. Liberty requires that the government not impose unreasonable restraints on the people and justice requires that the government treat like persons alike and not draw arbitrary distinctions based on irrelevant differences. Constitutional scholars have categorized constitutions and the rights they grant into three generations based upon the type of rights they grant.[46] The United States Constitution falls within the first generation and grants the people negative rights – rights to be free from governmental interference with the interest defined, such as speech, religion, or equality. More modern constitutions often grant positive rights and thus fall into the second or third generations. Second-generational rights involve rights of social welfare, such as the right of an individual to attain certain goods; for example housing, employment, minimum income, or medical care. Third-generational rights involve group or collective rights to interests that all people share in common, such as a clean environment and peace. However, the ability to enforce second- and third-generational rights is limited and as a result they often reflect merely the aspirations of the sovereign.[47] It is within the third generation of rights that some constitutions have begun to include animals.

It is difficult to amend the United States Constitution, requiring super-majorities and a deliberative process. The reason being that since the constitution is the highest law of the land and defines fundamental principles, the people should be able to rely on its stability and temporary majorities should not be able to pass amendments that weaken these principles. As a result, the United States Constitution has been amended only 27 times since 1791.[48]

In contrast, state constitutions are amended more easily. In fact, only 19 states have their original constitutions, and most have adopted three or more. Collectively, they have added over 5,000 amendments.[49] Each state defines the procedure for amendment.[50] Sixteen states allow amendment by popular initiative and referendum (I&R). I&R grants citizens the power to rule directly. While a republican form of government is the leading form nationally and within the states, the I&R is designed to enhance a republican form of government by strengthening the checks and balances on the state level, "to ensure that elected officials remain accountable to the electorate."[51] Generally, an initiative is proposed and if enough citizens sign the initiative, it is placed on the ballot for popular vote.[52] Given the malleability of state constitutions, animal advocates have used the voter initiative process to gain victories for animals by appealing directly to the voters rather than lobbying the legislatures. What animal advocates have found is that while winning the majority of voters is still difficult, they have a better chance of getting "what they want without any major concessions or compromises"[53] that are often necessary when lobbying politicians. According to one observer:

> Animal advocates have resorted to the initiative process after policy makers stonewalled popular reforms. For them, the initiative process has served as a safety valve, allowing popular sentiment to prevail over lawmakers bent on perpetuating the status quo.[54]

The Lisbon Treaty, arguably the functional equivalent of a constitution for the EU,[55] amended the Treaty on the European Union and the Treaty establishing the European Community, was ratified on December 1, 2009. The Lisbon Treaty expressly recognizes animals as sentient beings and requires that member states pay full regard to animal welfare when enacting laws. [56]

Public and private: statutory, regulatory and common law

Law is divided into two categories, public and private. The primary purpose of public law is to protect the public interest and welfare. Public law – criminal and civil – is designed either to incentivize conduct the society deems beneficial or prohibit conduct society deems detrimental. The purpose of the criminal law is to deter conduct that is deemed immoral and generally harmful and punish those who engage in such conduct. A criminal sanction involves either imprisonment or a monetary fine, set at a value to punish and deter. Since the criminal law is designed to punish and deter immoral conduct, and the sanction is extreme, the intent of the individual violating the law is a primary element of the crime.

Laws prohibiting animal cruelty, including animal fighting, abuse and neglect, are criminal laws. Because animals are "things" under the law, originally injury to an animal was found to cause harm only if such injury harmed the human owner. Thus, it was only a crime for a person to injure or kill an animal owned by another. The owner, however, was not restricted in his or her treatment of the animal. Over time the law expanded to prohibit cruelty inflicted on an animal by the owner. The justifications for this extension are that humans have a moral duty to the animals themselves and that animal cruelty is harmful to all humans as it might lead to violence against humans.[57]

The criminal laws are enforced by the police and prosecutors, although there are a very few examples of states that have granted standing to private entities to civilly enforce the criminal law. These statutes are especially important in the animal law context and provide greater opportunity to enforce the anti-cruelty laws.

A second function of the public law is to protect the public interest – health, safety and welfare – by regulating commercial and private conduct. Thus, an entire regulatory state has developed to protect humans as consumers from unsafe products, or as financial investors from misleading entrepreneurs, or as employees from unscrupulous employers, or as living beings dependent upon a clean and healthy environment. Two types of regulations concerning animals have been developed within this context: laws governing animal welfare and laws governing animal management and control.

The United States federal government enacted the Animal Welfare Act (AWA)[58] originally to protect pet owners from the theft of their animals for sale to research institutions by establishing licensing and other regulations for the sale of animals to such institutions. The AWA also sets minimum standards for the treatment of animals during transport and when used in specific contexts; for example, when bred for sale as companion animals, used by research institutions for testing and research, or used for certain types of public entertainment.[59] Another example of animal welfare legislation is the Humane Methods of Slaughter Act,[60] enacted to protect the health of consumers from adulterated food and to avoid the unnecessary suffering of animals during slaughter. These statutes provide for agency regulations to implement their goals. The regulations are drafted and enforced by the United States Department of Agriculture. They do not allow for private enforcement.

Additionally, the federal government, in order to protect the environment, manage wildlife, and comply with international treaty obligations, has enacted wildlife laws protecting certain classes of wildlife. The Endangered Species Act (ESA),[61] perhaps the most strict and protective of these laws, is designed to protect species from extinction and maintain an ecologically sound environment for humans. The regulations are drafted and enforced by the United States Fish and Wildlife Service, within the Department of the Interior. Private entities also may enforce the ESA.

Further, each state has enacted laws to "control" animals in order to protect public health and safety from animals deemed "vicious" or "pests." The laws are often drafted by the health authorities within a state and enforced by animal control officers or the local police. These laws provide for licensing and vaccinations of animals by their owners, define and regulate "dangerous dogs," manage free-roaming cats, and establish a bailee for people's lost (or abandoned) property, the government-owned animal shelter. States also have authority over fishing, hunting and trapping within their boundaries and thus enact laws to regulate humans' use of wild animals within the state. These laws are designed to manage state fish and game resources.

In the United Kingdom, the Department for Environment, Food and Rural Affairs (DEFRA)[62] is responsible for policy and regulations involving all animals, except those used in research. DEFRA's primary responsibilities include "The Environment,"[63] "Food and Farming,"[64] and "Wildlife and Pets."[65] Regulations governing animal cruelty, animal welfare, and the protection of wildlife fall within the jurisdiction of DEFRA. Additionally, DEFRA has enacted codes of practice governing a variety of animal communities, including dogs, cats, horses, privately kept nonhuman primates, and all animals used for food, including laying hens, meat chickens and breeding chickens, cattle, sheep, pigs and rabbits. The codes are not directly enforceable; however, a person's failure to comply may be used as evidence of guilt if that person is charged under the welfare regulations.[66] Such codes are drafted by Animal Welfare Councils, independent advisory bodies tasked with researching and developing animal welfare for each relevant animal community.[67] Animals used in research are regulated both by the Home Office for Science Research and Statistics[68] pursuant to the Animals (Scientific Procedures) Act 1986 and the local jurisdictions.[69]

Public decision-making in the EU involves several entities including the European Commission, the European Parliament, and the Council of the European Union. In general, the European Commission proposes new legislation, but the Council and Parliament enact the laws.[70] The laws of the EU are divided among regulations, directives, decisions, recommendations and opinions. Article 288 of the consolidated treaties defines each source as follows:

> A regulation shall have general application. It shall be binding in its entirety and directly applicable in all Member States.
>
> A directive shall be binding, as to the result to be achieved, upon each Member State to which it is addressed, but shall leave to the national authorities the choice of form and methods.
>
> A decision shall be binding in its entirety. A decision which specifies those to whom it is addressed shall be binding only on them.
>
> Recommendations and opinions shall have no binding force.[71]

These sources are described as "secondary legislation, the third major source of Community law after the treaties (primary legislation) and international agreements."[72] The EU has promulgated a number of regulations and directives that govern animal welfare for domestic animals, including companion animals, animals used for food, animals used in research, and animals in zoos.[73]

Finally, private law defines and facilitates interpersonal relationships and includes the substantive areas of torts, contracts, property, and family law. The private law defines the rights and obligations of humans vis-a-vis other humans and regulates human conduct by setting norms to protect citizens from harm. The conduct prohibited generally is not considered immoral but harm is nevertheless caused. The primary function of the private law is remedial. The law grants a right of action to a person harmed by another to seek compensation or other relief from the person that caused the harm. Property law allows individuals to bring claims against those who take or destroy their property. Tort law establishes minimum standards for reasonable conduct and creates private claims for persons injured by the unreasonable or intentional acts of another. Contract law allows individuals to create rights and obligations through agreements with others and provides for the enforcement of those agreements in a court of law. Family law provides a legal structure for the creation and dissolution of a marriage or other civilly recognized intimate partnership and the incidents that arise within such marriage or intimate partnership. Trust law allows for a person to place in trust resources for the care and benefit of themselves or others.

Since animals are owned by humans as property, the private law involving animals functions to protect human interests in their animals. The private law governs issues such as the ability to transfer ownership of an animal, to seek compensation for the economic loss due to the wrongful injury or death of an animal by another, to hold owners of animals responsible for harm their animals cause to other humans, to award custody of the companion animal to a person when partners break up, and to provide for the care of a companion animal upon the incapacity or death of the human owner.

Legal rules and legal rights

Legal rules encourage, allow, or prohibit actions. Legal rules fall into two general categories: rules of reason (standards) and per se (bright-line) rules. Rules of reason are written in a manner that expressly provides for individual discretion. For example, a rule may prohibit acts that cause an animal *unnecessary* pain or suffering. This is an example of a rule of reason because the actor must use discretion to determine what is *necessary* and thus what constitutes compliance with the rules. By definition, compliance with a rule of reason requires individual balancing on a case-by-case basis. Per se rules are rules that are drafted in a manner that allow for little, if any, discretion. For example, a rule that prohibits acts that cause an animal pain, is a per se rule. No discretion is necessary to determine what acts are allowed, if they cause the animal to feel pain, independent of the individual circumstances, they are prohibited. Note that the terms used in a rule, for example the term "pain," may themselves be open to interpretation. Thus, there may be some level of discretion in determining when an animal feels pain. However, the rule itself is not drafted in a manner deliberately to grant discretion.

What does it mean to have a legal "right"? A legal right is a right granted by the law, in contrast to moral or natural rights derived from philosophical, religious or other foundational principles. Wesley Newcomb Hohfeld, a renowned American professor of jurisprudence, distinguished among a variety of legal interests, often referred to as rights, and determined that only claim rights are true rights.[74] A legal claim right exists when the law grants an individual, P, a substantive guarantee of some sort, either immunity from harm or grant of a positive benefit, and imposes a duty upon another, D, to either refrain from harming P, or to benefit P. Thus, P has a claim against D if D fails to comply with his or her duty. Moreover, a legal right includes the ability to enforce the substantive guarantee, publicly and/or privately, and obtain a remedy. Claim rights are distinguished from what some refer to as liberty rights, or, as Hohfeld considered, a privilege in which the law grants P the freedom to act but D has no corresponding duty to P.[75] For example, many anti-cruelty laws protect animals from neglect by their owners. Such laws require that the owner provides his or her animal with adequate food, water and shelter. These are criminal laws in the United States and thus enforceable by the police or humane law enforcement officers. These laws arguably grant animals legal rights because they define a substantive guarantee to adequate food, water and shelter; impose a duty on the human owner to provide adequate food, water and shelter; are enforceable publicly; and are remedied by seizure of the animal (allegedly to a shelter where the animal will receive adequate care) if the owner does not comply with his or her duty.

Animals as things owned as property

The basic construct of property

The law in most all countries characterizes animals as "things" who are owned as personal property. The legal construct of "property" and the related concept of "ownership" are very broad and abstract. The Roman law divided all things into two classes – things private and those which are common.[76] Wild animals fell into the common class, meaning they belonged in common to all citizens. Under natural law, one gained ownership over things through dominion. This rule is called the "animals ferae naturae doctrine" and provides that wild animals reside within the common and belong to no one as long as the animal remains wild, unconfined, and undomesticated. Sir William Blackstone, an English judge who wrote a treatise on the common law that remains an important source for the classical view, commented:

> A man may lastly have a qualified property in animals ferae naturae-propter privilegium; that is, he may have the privilege of hunting, taking and killing them in exclusion of other persons. Here he has a transient property in these animals usually called 'game' so long as they continue within his liberty, and he may restrain any stranger from taking them therein; but, the instant they depart into another liberty, this qualified property ceases.[77]

Thus, in order for an individual to claim ownership of a wild animal, the individual must have actual possession, custody, or control by taming, domesticating, confining or killing the wild animal. Moreover, the mere presence of a wild animal on a landowner's real property does not grant ownership rights of the landowner to the wild animal.

Since wild game resides within the common for the benefit of the people, the sovereign retains title to all wild game in trust for the benefit of the public. The sovereign, in holding title, has the power to enact laws to manage or preserve the game for the common good. This understanding was summarized by the Napoleonic Code which read: "There are things which belong to no one, and the use of which is common to all. Police regulations direct the manner in which they may be enjoyed. The faculty of hunting and fishing is also regulated by special laws."[78]

In the United States, the authority to regulate and control wild animals resides with the state governments.[79] Of course, the state must act in accordance with the United States Constitution. Thus, where the Constitution grants power to the federal government, as in the power to enter into treaties with other countries, the federal government will also have power to regulate and control wild animals.

Types of property

The law of property is divided into four groups each reflecting a different type of thing that is the object of ownership: real property – land and fixtures affixed to the land; personal property – movables or chattels; intangible property – intellectual property such as copyrights, patents, and trademarks; and cultural property – precious, irreplaceable items of significance to a cultural group such as antiquities, national monuments, and folklore. Animals, when owned, fall within the class of personal property.

What does it mean to have a property right or ownership in a thing? The most traditional characteristic of ownership of an object is one's right to exclude another's use of the object. However, this is not the only characteristic of ownership. Property rights are often referred to as a "'bundle of rights' that may be exercised with respect to the object – principally the rights to possess the property, to use the property, to exclude others from the property, and to dispose of the property by sale or by gift."[80] All rights within the bundle may not apply to all forms of property. Moreover, the concept of property invokes not only privileges but also obligations as the law may impose restrictions on one's use for a variety of policy reasons. Thus, for example, zoning, nuisance or historical preservation laws may restrict one's ownership interest in land. Public health and safety laws restrict the sale and use of certain items of personal property, such as food, drugs, firearms, waste products, and animals, and regulate the operation of vehicles.[81] Nevertheless, to the extent one retains at least one stick of the bundle, one is deemed to have a property interest in the object. Thus, ownership of an animal means that the owner has (1) the right to possess, use, transfer, dispose of, and exclude others from taking the animal, and (2) the obligation to the animal, if defined by law, such as the duty to provide adequate food, water and shelter.

Moving on from the basics

This very brief introduction provides a bird's-eye view of the remainder of this text. The following chapters, organized around the sources of law governing animals, will explore each in detail. Chapter 2 begins with an exploration of the anti-cruelty laws that provide the most fundamental and basic protection for animals within society. Next, Chapter 3 discusses the laws that regulate certain commercial uses of animals, the statutory and regulatory animal welfare laws. Chapter 4 then turns to the laws that address public health, safety and the environment, the animal control and management laws. Chapter 5 focuses on the two legal regimes in which animals generally are absent as subjects, the constitution and the private law. Although absent as subjects, these laws do protect human owners' interests in their animals and in turn affect animals' interests as well. The final chapter explores possibilities for the future of animal protection law by summarizing several of the legal theories proposed to improve upon the laws, analyzing how effective these theories are likely to be, and proposing a few basic principles to create an all-inclusive legal regime that recognizes, values and protects all animals' inherent interests in themselves as sentient beings, independent of their relationship to humans.

2
Anti-Cruelty Laws

The United States anti-cruelty laws are criminal laws and comprise the most basic and fundamental legal protection for animals. As the least controversial of the animal protection laws, these laws protect animals from the intentional and gratuitous infliction of pain and suffering at the hands of humans. However, as we will see, most animals in the United States are not covered by the anti-cruelty code. Animals used for food or research and wild animals generally are exempt from the anti-cruelty laws. The anti-cruelty laws target only individual instances of intentional cruelty not institutionalized cruelty. This reflects a serious flaw in our societal view of animals as beings who exist for human use and our acceptance of institutionalized practices that inflict serious pain and suffering upon other species for our benefit.

Pain and suffering are terms used in most anti-cruelty laws but are difficult to distinguish. For example, Washington state law provides:

> A person is guilty of animal cruelty in the first degree when, except as authorized by law, he or she, with criminal negligence, starves, dehydrates, or suffocates an animal and as a result causes: (a) substantial and unjustifiable physical *pain* that extends for a period sufficient to cause considerable *suffering*; or (b) death.[1]

Thus,

> pain usually refers to the reaction following an adverse physical stimuli ... to be an unpleasant sensation ... Suffering is sometimes taken to be pain of a certain sort – for example, sufficiently unwarranted, prolonged, and outside the control of the subject. However, while pain usually accompanies suffering, it is not always identical to it. Suffering thus refers to physical pain, including what has been termed

the mental experience of pain, including such sensations as shock, fear, foreboding, anxiety, trauma, anticipation, stress, distress, and terror. In general, suffering may be defined as harm that an animal experiences characterized as a deficiency in (or negative aspect of) that animal's well-being.[2]

State statutes

The basic elements

In the United States, animal cruelty is governed by state law and thus each state and the District of Columbia have different definitions of what constitutes actionable animal cruelty. Two states – Florida and Illinois – have been selected as contrasting examples. We will compare and contrast their approaches to criminalizing animal cruelty and reflect on which approach provides greater protection and why.

Florida law defines animal very broadly, including "every living dumb creature." The law also defines "torture," "torment," and "cruelty" "to include every act, omission, or neglect whereby unnecessary or unjustifiable pain or suffering is caused, except when done in the interest of medical science, permitted, or allowed to continue when there is reasonable remedy or relief."

Florida law establishes a minimal duty of care to animals by sanctioning whoever:

(2) (a) Impounds or confines any animal in any place and fails to supply the animal during such confinement with a sufficient quantity of good and wholesome food and water,

(b) Keeps any animals in any enclosure without wholesome exercise and change of air,

or

(c) Abandons [e.g. forsakes an animal entirely or neglecting or refusing to provide or perform the legal obligations for care and support of an animal by its owner] to die any animal that is maimed, sick, infirm, or diseased.[3]

Additionally, Florida defines "Cruelty to Animals" as:

(1) A person who unnecessarily overloads, overdrives, torments, deprives of necessary sustenance or shelter, or unnecessarily mutilates, or kills any animal, or causes the same to be done, or carries in or upon

any vehicle, or otherwise, any animal in a cruel or inhumane manner, is guilty of a misdemeanor of the first degree ...

(2) A person who intentionally commits an act to any animal which results in the cruel death, or excessive or repeated infliction of unnecessary pain or suffering, or causes the same to be done, is guilty of a felony of the third degree, ...

(a) A person convicted of a violation of this subsection, where the finder of fact determines that the violation includes the knowing and intentional torture or torment of an animal that injures, mutilates, or kills the animal, shall be ordered to pay a minimum mandatory fine of $2,500 and undergo psychological counseling or complete an anger management treatment program.[4]

Florida expressly exempts veterinary practices, research on animals, and traditional animal husbandry practices from the law.

Compare Florida with the Illinois law entitled the "Humane Care for Animals Act." The Act defines animals very broadly as well: "every living creature, domestic or wild, but does not include man."[5] Illinois imposes an affirmative duty of care upon owners[6] to provide the basic necessities of life to their animal, including "(a) sufficient quantity of good quality, wholesome food and water; (b) adequate shelter and protection from the weather; (c) veterinary care when needed to prevent suffering; and (d) humane care and treatment."[7] The violation of this provision constitutes neglect of the animal and is a misdemeanor with a sentence of imprisonment for not more than six months.

Illinois' anti-cruelty statute prohibits the following acts by any person:[8]

- *Cruel treatment*, which includes beating, cruelly treating, tormenting, starving, overworking, abandoning, or otherwise abusing any animal.[9]
- *Aggravated cruelty*, which is defined as intentionally committing "an act that causes a companion animal[10] to suffer serious injury or death ... [except] euthanasia ..."[11]
- *Torture*, which means "infliction of or subjection to extreme physical pain, motivated by an intent to increase or prolong the pain, suffering, or agony of the animal" unless legally justified.[12]

Legal justification for torture includes hunting, fishing, trapping, and practices by veterinarians. Moreover, normal husbandry practices are exempt from the entire cruelty code. Thus, it is perfectly lawful under Illinois law to torture wild animals and animals used for food.

As with all criminal laws, some level of knowledge or intent is required to justify sanction, and the more culpable the intent, the greater the justification for sanction. Thus, if a person intends, knows, or should have known that he or she was subjecting the animal to cruel treatment that person is subject to a Class A misdemeanor with a sentence not to exceed one year. Aggravated cruelty requires the intent to commit an act that causes a companion animal to suffer serious injury. As such, it is a Class 4 felony with a sentence of imprisonment for not less than one year and not more than three years. Finally, the act of torturing an animal requires that the actor be motivated by the intent to cause the animal extreme pain and suffering and is a Class 3 felony with a sentence of imprisonment for not less than two years and not more than five years. Aggravated cruelty is a crime only when inflicted on a companion animal. This, in turn, suggests that elevated sanctions are warranted only when the human owner has a personal attachment to the injured animal.

The statute further provides that in addition to any other penalty, the court may order the convicted person to undergo a psychological or psychiatric evaluation at the person's expense. Moreover, if the person is a juvenile or companion animal hoarder the court must order such an examination. The statute defines "Companion Animal Hoarder" as a

person who (i) possesses a large number of companion animals; (ii) fails to or is unable to provide what he or she is required to provide under Section 3 of this Act; (iii) keeps the companion animals in a severely overcrowded environment; and (iv) displays an inability to recognize or understand the nature of or has a reckless disregard for the conditions under which the companion animals are living and the deleterious impact they have on the companion animals' and owner's health and well-being.[13]

Hoarding[14] is a very serious and unique form of animal cruelty and neglect that often involves a person who believes he or she is helping the animals but instead he/she is severely neglecting them. Hoarders may have hundreds of cats in their home, many who may be dead, and yet they continue to take in more animals. Because the conduct is often a result of a psychological disorder,[15] the Illinois law recognizes the need for psychological monitoring and requires that the court order psychological counseling.

How do the two sample states compare? Which state provides the better protection for the animals? The most striking difference appears in the definition of cruelty. Florida expressly defines "cruelty" and "torture" as "every act, omission, or neglect whereby *unnecessary or unjustifiable* pain or suffering is caused." The acts constituting cruelty include "*unnecessarily* overloads, overdrives, torments, deprives of necessary sustenance or shelter, or *unnecessarily* mutilates" an animal. In contrast, the words "unnecessary or unjustified" are absent from the Illinois statute when defining cruel treatment or aggravated cruelty. However, Illinois uses the term "cruel" without defining it. The standard definition of cruel is "to inflict pain and suffering," although the courts may consider the circumstances when determining guilt. The express use of the terms "unnecessary" or "unjustifiable" fundamentally alters the protection afforded animals under the law. This rule of reason allows for individual balancing to determine what is unnecessary or unjustified under the circumstances, and thus the definition of what constitutes cruelty is highly discretionary.

Illinois also appears to require a more robust duty of care. The Florida statute imposes the duty upon anyone who "impounds, confines, or keeps" an animal, whereas Illinois imposes the duty upon owners. The definition of "owner" under Illinois law includes anyone who keeps or harbors or has custody of an animal. Thus, although the scope of persons with a duty of care in both sample states is the same in practice, the duty imposed by Illinois law is slightly broader and includes an affirmative duty to provide humane care.

As for exemptions under the laws, both states exclude normal animal husbandry practices, thus providing little if any protection for animals used for food. Both states also exclude veterinary practices arguably to protect veterinarians from prosecution for inflicting pain when not intended. Florida expressly excludes research, while Illinois excludes hunting, fishing and trapping. These exemptions, and others, like exemptions for zoos or rodeos, are quite common among the states.

In comparison, the United Kingdom, which completely redrafted its animal welfare laws in 2006 to consolidate and modernize the existing laws at the time, makes it an offense to cause an animal to suffer unnecessarily. The Animal Welfare Act, chapter 45, section 4(3), lists the factors that should be taken into account in determining whether the suffering is unnecessary:

(a)   whether the suffering could reasonably have been avoided or reduced;
(b)   whether the conduct which caused the suffering was in compliance with any relevant enactment or any relevant provisions of a licence or code of practice issued under an enactment;

(c) whether the conduct which caused the suffering was for a legitimate purpose, such as –
   (i) the purpose of benefiting the animal, or
   (ii) the purpose of protecting a person, property or another animal;
(d) whether the suffering was proportionate to the purpose of the conduct concerned;
(e) whether the conduct concerned was in all the circumstances that of a reasonably competent and humane person.

This list of factors is useful to provide a meaningful limitation on the interpretation of "unnecessary." Note that justification involves both the purpose of the activity as well as the methods used to treat the animal. It is expressly noted that the animals' suffering should be in proportion to the purpose and, when reasonable, the suffering should be reduced or avoided. Arguably, when an accommodation that would avoid or reduce the level of suffering the animal endures involves only minimal expense to the owner, it should be considered "reasonable." In these circumstances, failure to make the accommodation should be considered a violation of the Act.

The Act also requires one who is responsible for an animal to take all reasonable steps to ensure the animal's needs are met. Those needs include a suitable environment, suitable diet, ability to exhibit normal behavior patterns, to be housed with, or apart from, other animals, and protected from pain, suffering, injury and disease, taking into account the lawful purpose for which the animal is kept and the lawful activity undertaken in relation to the animal.[16] This provision was added in 2006 and represents the first time the law in the United Kingdom imposed a duty of care upon owners of domestic animals rather than merely protecting animals from individual intentional cruelty.

In New Zealand, the Animal Welfare Act of 1999[17] Section 29(a) makes it an offense to "ill-treat" an animal, which in turn is defined in section 2 as "causing the animal to suffer, by any act or omission, pain or distress that in its kind or degree, or in its object, or in the circumstances in which it is inflicted, is unreasonable or unnecessary." In February 2010, New Zealand introduced legislation to distinguish willful ill-treatment from reckless ill-treatment, where a person knew serious harm to an animal would occur and unreasonably risked the harm.[18] The bill states: "These amendments correspondingly respond to an increase in the number of serious animal welfare incidents being investigated and will send a strong signal to the judiciary that the Government and general public wish to see heavier penalties for this type of offending. The increased penalties will also act as a deterrent to potential offenders."[19]

The Act states that the pain and suffering inflicted must be unreasonable or unjustified. This invokes a utilitarian balancing of interests – the benefit or gain to the human against the cost to the animal. The first step for the court is to establish the purpose for inflicting the pain. Inflicting pain for no legitimate reason is prima facie evidence of unnecessary suffering. However, if there is some legitimate reason or purpose for inflicting pain, the court then must determine if the means used are legitimate. As Australian academic Peter Sankoff noted, "the list of 'legitimate' uses is virtually endless,"[20] requiring only that the use be an accepted part of human privilege that advances society's interests. If the pain is imposed to satisfy human hunger, enjoyment, or even aesthetic preference (as in docking a dog's tail), it is deemed legitimate. The legitimacy of the means used is determined based on efficiency. Only if a less painful procedure is available at comparable cost will the means utilized be deemed not legitimate.[21] Note that at every stage in the calculus, human interests trump the animal's interest.

Finally, as mentioned above, Illinois exempts "normal husbandry practices" from the entire cruelty code and "hunting, fishing, trapping and practices by veterinarians" from the definition of torture. Florida also exempts scientific research. These exemptions are typical of most state anti-cruelty laws. The result is that the criminal laws, designed to protect animals from the intentional infliction of pain and suffering, perpetuate and in fact endorse institutionalized cruelty to animals.

In contrast, the United Kingdom cruelty laws do *not* exempt animals used for food or accepted agricultural practices. Note, however, that in listing the factors for determining "unnecessary suffering," actions that are in compliance with codes of practice or other laws are considered. The only express exemption in the United Kingdom is for animals used in research. A separate Act governs these animals.

The link

Important to the development of the anti-cruelty laws is the well-established link between individualized acts of animal abuse and human violence. Studies have shown that animal abuse is both a predictor of domestic violence and an indicator of co-occurring domestic violence.[22] In particular, there is clear evidence that in homes where there is domestic violence/spousal abuse, there is a higher incidence of animal abuse.[23] Whether their victims are human or nonhuman, the abuser abuses a weaker family member to assert power and control and to intimidate.[24]

The abuse is *family* abuse and all members of the family are affected, both physically and emotionally.[25] Moreover, children exposed to family violence are at a greater risk of emotional and behavioral problems, very often including abusing animals.[26] Childhood animal abuse is linked to other forms of aggressive behavior in children, juvenile delinquency, and criminal activity as an adult, including violent behavior.[27] In fact, animal cruelty is one of four factors that predict interpersonal violence.[28] Thus, a child in an abusive family environment is more likely to become an abuser as an adult, creating another family of abuse. The result is a cycle of violence that is complete and continues.

Further, outside the family context, studies have demonstrated a predictive link between those who deliberately and violently kill cats and dogs and those who engage in the serial killing of humans.[29] Often, the individual begins at an early age abusing animals and then progresses over time to violently harm humans. For example, "as a boy Jeffrey Dahmer impaled the heads of cats and dogs on sticks; Theodore Bundy, implicated in the murders of some three dozen people, told of watching his grandfather torture animals; and David Berkowitz, the 'Son of Sam,' poisoned his mother's parakeet."[30]

Researchers into the link recognize that culturally and legally we draw distinctions between individualized animal cruelty to companion animals and institutionalized cruelty against other animals. Heide and Merz-Perez describe this state of affairs as a "contradiction ... in human behavior" but, when "properly conceptualized," may be understood.[31] They cite the classic example of the criminal who is incarcerated for murder for killing a person while a soldier in battle is heralded as a hero for killing a person. The act of killing another human itself is not necessarily wrong or evil. What society deems unacceptable is to engage in such conduct for an improper reason; in other words, unjustified. The researchers explain that "no society can survive without reconciling the many contradictions posed by human nature," at least intellectually, if not necessarily ethically.[32]

Within the context of animal cruelty and the link to human violence this reconciliation is achieved by identifying and emphasizing the importance of the context and the motive behind the cruelty. For example, society draws a distinction between a farm boy who finds a mouse in the feed and kills the mouse and a city boy who kills his sister's pet mouse. The farm boy's act of killing the mouse is deemed acceptable because it can be justified as eliminating a pest that might contaminate the feed, unless perhaps the farm boy tortured the mouse in the process. On the other hand, the city boy's killing of the mouse that is loved by his sister is unjustified and deemed unacceptable and even evil because the animal was loved by his sister and the killing did not serve any legitimate purpose. However, while such reconciliation may be important to the research on the link between currently defined cruelty against animals and criminal human violence, it offers little insight into why we not only justify but accept and promote many acts of institutionalized cruelty against animals. Moreover, this reconciliation further demonstrates that society is concerned with harm to the human but not to the animal. The mouse, whether on the farm or in the city, is the same animal with the same interests. The justification for accepting the killing of the mouse on the farm is because in that scenario a human deems the mouse a "pest," while the killing of the city mouse is deemed unjustified solely because it inflicts harm on the human owner. Thus, the value a human places on an animal will determine whether that animal has a right to live. Clearly lacking is any discussion of the harm to the mouse as a sentient being with inherent interests.

Recently, researchers have begun to study the correlation between the acts of slaughtering animals in a slaughterhouse and the increased crime rates in the surrounding community.[33] Theoreticians have linked increased crime rates in these communities to a number of variables, including "demographic characteristics of the workers, social disorganization in the communities, and increased unemployment rates." However, they have not been empirically tested. In 2009, three researchers published the findings of a study to test the following general hypothesis:

> Controlling for the variables commonly proposed to explain crime, slaughterhouse presence and employment will be associated with increased crime rates. These increases will be greater than those observed from industries that use the same type of labor force, have high injury and illness rates, and entail routinized labor, but do not involve killing and dismembering animals. In particular, rape and family violence will be influenced by slaughterhouse work, net of other factors.[34]

The study supported the hypothesis. After controlling for key variables that have been linked to increased crime rates in a community – the number of young men, population density, the total number of males, the number of people in poverty, international migration, internal migration, total non-White and/or Hispanic population, unemployment rate, and the total population of the community – the results demonstrated that "slaughterhouse employment is a significant predictor of both arrest and report rate scales ... [and] has significant effects on arrests for rape and arrests for sex offenses."[35] This study is an important first step to understanding the link between institutionalized cruelty and violence against humans. As such research develops, the findings may justify a dramatic shift away from the currently socially accepted institutionalized human cruelty against animals; if not to benefit the animals, at least to benefit humans.

Interestingly, two state supreme courts have already noted the relevance of slaughterhouse killing in the prosecution of individuals charged with murder. The Florida Supreme Court upheld against challenge the state's slaughterhouse arguments during closing when the prosecutor analogized the defendant's conduct to his use of knives to kill animals in the slaughterhouse.[36] The court held that the evidence based on the defendant's testimony describing in detail his slaughtering of cows, pigs, and hogs, supported the analogy. Similarly, the California Supreme Court upheld against challenge the admission of evidence of the defendant's slaughterhouse employment to show that the defendant treated the victim more like an animal than a human.[37] The court found that the evidence describing the killing of animals in the slaughterhouse was sufficiently similar to the pathologist's testimony describing the manner in which the victim was murdered to make the evidence relevant and not unduly prejudicial.

The link between animal cruelty and human violence incentivizes the enforcement and prosecution of anti-cruelty laws. Moreover, it suggests ways in which the law may target and capitalize upon the related abuses rather than only addressing each type of abuse independently – domestic, child, elder and animal – to better protect all victims. For example, consider a "cycle of laws" that might address the "cycle of violence" noted by the researchers.[38] At every step along the cycle – from before the abuse begins, through sanction and rehabilitation after the abuser is found guilty – the law may properly reflect the link.

- Step 1: Avoid all abuse by including humane education programs in schools to teach young children respect and compassion for all living sentient beings.
- Step 2: Prevent human abuse through the aggressive enforcement of effective anti-cruelty laws that properly reflect the nature of the crime through sentences designed to deter, punish and rehabilitate the offender.
- Step 3: Detect abuse earlier by maintaining databases that track animal cruelty and other violent crimes separately and mandating

the direct and cross-reporting of suspected abuse by proper authorities; for example, physicians, veterinarians, and teachers, and related agencies, such as humane law enforcement and child and adult protective services.

- Step 4: Enhance protection of all victims of family abuse by facilitating the early securement of a protective order upon a showing of animal abuse in the home, giving express authority to the courts to issue protective orders that direct the care, custody, or control of a companion animal residing in the home, and providing safe havens for all victims of abuse, including the animals.
- Step 5: Facilitate the prosecution of abuse by recognizing the relevance of connected acts of abuse within the home and allowing the introduction of evidence of such acts in the prosecution and trial of the abuser.
- Step 6: Avoid further abuse through appropriate sanction of the convicted abuser by prohibiting his or her ownership of animals for an appropriate period of time, imposing appropriate sanctions, and requiring proper psychological treatment to facilitate rehabilitation of the accused offender.

A progressive anti-cruelty law: Puerto Rico

In August 2008, Puerto Rico[39] enacted the Animal Protection and Welfare Act[40] that distinguished Puerto Rico among the more progressive jurisdictions in its treatment of animal cruelty. The legislature opened the Act with a "Statement of Motives" that acknowledges that:

> animals are sensitive beings that are entitled to humane treatment ... [and] that deserve being treated with fairness and dignity ... Puerto Rico should rise above as a sensitive and progressive society that respects, protects and cares for its animals. A new Act is necessary not only for the protection of these defenseless beings, but also to collaborate in forging a Puerto Rican society that is mentally healthy.

What makes the Puerto Rican law progressive is not only that the legislature is recognizing the inherent interests of its animals, but that it addresses many of the deficiencies of other anti-cruelty laws. For example, there are no exemptions for any captive animal, including animals used for food or protected wildlife.[41] With respect to experiments with live animals, only experiments "absolutely essential for scientific research purposes at university centers" are permitted.[42] Section 2(f) defines a standard of "minimal care" that is higher than most, by requiring "care to preserve the health and welfare of an animal." Such care includes:

> quantity and quality of enough food to allow for the growth or maintenance of the normal body weight of the animal; open and proper access to drinking water, at a temperature that is fit for drinking and in sufficient amount to satisfy the needs of the animal; ... structure able to protect the animal from bad weather conditions; ... veterinary care that a prudent person would deem necessary to protect the animal from suffering, [including] vaccination and preventive care; ... [and] continuous access to an area [for exercise].

The law protects an animal from unnecessary suffering, defined in Section 2(u) as "suffering that is not necessary to ensure the safety, health, or welfare of the animal or other beings within its environment." Thus, the suffering inflicted is necessary only when promoting the safety, health and welfare of the animal or other beings. The criminal sanctions are progressive, focusing both on the level of egregiousness of the defendant's conduct and state of mind as well as the level of harm inflicted on the animal. The criminal sanctions include felony sanctions, with minimum as well as maximum sentences, that better reflect the seriousness of the crimes. Finally, the law demands enforcement and requires coordination with organizations that provide services to animals. Section 3 states that "municipalities ... shall assign top priority to handling situations that come to their attention and which involve abuse and/or negligence of stray animals." Ironically, cock-fighting remains legal in Puerto Rico although outlawed in all 50 states and the District of Columbia.

Common law: interpreting and applying anti-cruelty statutes

Analysis of cases applying the law is critical to understanding how the law is interpreted and enforced. For example, an interesting case from 1975 involved a horse named Mabel who pulled a hansom cab.[43] After an inspector noticed that Mabel was limping, he notified the driver. Although the driver suspended Mabel's service for that day, he had her pulling the cab the next day while still limping. The inspector issued a summons to the driver for violation of the animal cruelty law.[44] Three issues were presented to the court: "Are omission and neglect punishable ... as are incidents of active cruelty? Does driving a lame horse constitute torture? Is the element of a culpable state of mind necessary for conviction, and if so, have the People demonstrated sufficiently that such state of mind existed here?"[45]

The court noted that the test of cruelty is the justifiability of the act or omission and that working an animal unfit for labor could constitute an act of cruelty. The court questioned whether the failure to provide medical care would also qualify as cruelty under the provision "deprives any animal of necessary sustenance, food or drink." The court relied on Webster's Dictionary definition of "sustenance" as "the necessities of life" and found that medical care when ill falls within the definition. Interestingly, in 2008 a New York court interpreting the same statute held that

> [a] plain reading of the statute reveals that "necessary sustenance" is described within that clause as "food or drink." The grammatical construction of the clause "or deprives any animal of necessary sustenance, food or drink, or neglects or refuses to furnish it such sustenance or drink" indicates that "necessary sustenance" is "food or drink" and thus did not include medical care.[46]

Regarding the mens rea, or culpable state of mind, of the defendant, the court explained that he need not maliciously hurt the horse; in other words, work the horse with the purpose of torturing the animal. "However, if the defendants knew about the horse's condition, and the horse suffered torture or pain when it was driven, they must be presumed to have known that if they caused the horse to be worked it would suffer. The question is whether they willfully caused certain things to be done, i.e., driving the horse, which necessarily tortured it."[47] In this case, the defendant had been told by the inspector that Mabel was limping. Moreover, Mabel was seen by a vet two weeks after the summons issued. The vet testified at trial that the examination suggested that the problems had developed over an extended period of time. Further, although Mabel could not verbalize her suffering, her limping demonstrated that she suffered pain from having to pull the cab while lame. The judge found the evidence supported a conviction.

Perhaps most interesting are the following comments of the judge:

> [T]he moral obligation of man toward the domestic animal is well documented in the Bible. "A righteous man regardeth the life of his beast" (Proverbs 12:10). He has consideration for its feelings and needs. The Bible also states that if you see an animal hurt or overburdened, one should not look away but help it (Deuteronomy 22:4.) It is truly a humanitarian sentiment that domestic animals are in fact considered

part of the human community. Thus, they should be treated with respect and given proper care.[48]

Reference to the scriptures is unusual in animal cruelty cases. This reference suggests and supports the moral underpinning of the anti-cruelty laws, and the purpose not only to prohibit cruelty to protect the animal from pain and suffering but to ensure a "righteous" society.

Targeting animal fighting

All states and the federal government outlaw animal fighting. Animal fighting is a special case of cruelty where owners force their animal to fight another animal, often to the death. The two most common types are cock-fighting[49] and dog-fighting.[50] The animals raised for fighting are often injected with hormones and tormented, and the dogs are often baited with live animals such as dogs, cats and rabbits (often stolen from their owners) to make them more aggressive. Of course the pain and suffering inflicted on the bait-animals is quite great as well. Moreover, the fighting animals are forced to continue to fight even after suffering severe injury. The losing animal, if not killed during the match, often is killed afterward by the owner simply because he or she lost. Paraphernalia, such as blades that are attached to the birds' feet to inflict mortal wounds on the other bird during the fight, or treadmills and heavy chains used to "strengthen and train" dogs to fight, are common.

Animal fighting receives more attention from police and prosecutors than other animal cruelty crimes because those engaged in animal fighting also tend to be involved in additional criminal conduct, such as gambling, guns, and illegal drugs. In 2007, when Michael Vick, a star American football player in the National Football League, was arrested for animal fighting and ultimately convicted, the horrors of the cruelties involved in such enterprises were widely publicized and the public was outraged. Since then, much more attention has been focused on the stiffening and enforcement of animal fighting laws.

The state of Virginia now has a comprehensive animal fighting statute.[51] The statute criminalizes virtually all behavior connected with animal fighting, including promoting, engaging in, aiding or abetting, or even attending a fight. This broad targeting is crucial to effective enforcement because when an animal fighting ring is uncovered, proving each individual's participation at the event is very difficult. Thus, all participation is outlawed and individual proof of whether one was the ringleader or a spectator becomes less important.

Outlawing animal fighting is no longer controversial in the United States for the large majority of the population. Dog-fighting and cock-fighting are now illegal in every state.[52] Moreover, five states have outlawed the less common hog-dog fighting and the attorney generals of two states have stated that hog-dog fighting violates their anti-cruelty statutes.[53] However, there are people who believe that cock-fighting should be legal and that it is no worse than the treatment millions of birds receive on factory farms daily. Opponents of the laws banning cock-fighting have argued that gamecocks want to fight, it is a natural instinct (although in nature it rarely results in death to the losing cock), and it is a better fate than being raised for food and kept in a confined space before being slaughtered.

This raises an interesting ethical issue. Many would likely agree that the treatment of birds raised for food is inhumane, and perhaps, from the bird's point of view, not any worse than being forced to fight. However, a society's social and religious beliefs define the laws that govern their behavior. In the United States it is socially acceptable to allow birds raised for food to be inhumanely confined and otherwise abused; however, it is not socially acceptable to force birds to fight even though it may not be any more inhumane to the bird. Why is this so?

First, the enterprises serve two different functions and are conducted in very different places. Cock-fighting is a sport for human entertainment and thus performed in public, although clandestinely, watched by adults and children alike. Factory farming is a business operated for human consumption with operations hidden behind closed doors. Thus, except for the workers involved (recall the study cited previously suggesting a link between such work and increased violent crime), humans are not exposed to the cruelty of factory farming. Also, Americans enjoy eating chicken and eggs and want to do so at an inexpensive price. Thus, there is little incentive to regulate factory farms.

Further, most Americans do not view cock-fighting as a legitimate sport. In contrast, greyhound racing is a socially acceptable American sport, even though the dogs and many other small animals suffer severely in this enterprise as well. In fact, greyhound racing arguably is just as cruel to the dogs as is dog-fighting. Often, live lures are used to train greyhounds to run in a circle for the race by dangling small animals including jackrabbits, kittens, guinea pigs and chickens as live bait to entice the dogs to run. This technique is similar to the use of live bait to train dogs to fight. The cruelty to these animals in both settings, racing or fighting, is great. Moreover, thousands of dogs are seriously injured each year and thousands more are killed when no longer profitable,[54] much like dogs used to fight. Furthermore, the greyhounds are housed under conditions as inhumane as the chickens in factory farms. Commercial greyhound racetracks house thousands of greyhounds in warehouse-like kennels filled with small cages barely large enough for the dog to stand up. Such cruelty is allowed merely for human profit and the supposed pleasure of the viewing public. But is the life of a greyhound racer that much different from that of a fighting cock or dog?

Finally, the factory farms are huge employers and wield much political clout, while many of those engaged in cock-fighting tend to be marginalized minorities with cultural beliefs different from those of the American majority. Thus, from a human's perspective, factory farms and cock-fighting are completely distinct. However, from the animal's perspective, there is likely little difference – they suffer and are killed either way.

Given the widespread acceptance that animal fighting is wrong, the federal government also criminalizes animal fighting ventures that affect interstate commerce under the Animal Welfare Act. The relevant part states: "it shall be unlawful for any person to knowingly sponsor or exhibit an animal in an animal fighting venture [or] ... knowingly sell, buy, possess, train, transport, deliver, or receive any animal for purposes of having the animal participate in an animal fighting venture."[55] Since the animals, paraphernalia, or people involved in the enterprise often cross state lines, animal fighting is punishable under both state and federal law.

Similarly, the United Kingdom Animal Welfare Act of 2006 re-codified the long-standing law prohibiting animal fighting.[56] An animal fight is defined as "an occasion on which a protected animal is placed with an animal, or with a human, for the purpose of fighting, wrestling or baiting,"[57] where a "protected animal" is a vertebrate animal not in the wild. The offense includes all aspects of participating in an animal fight, including causing, publicizing, taking part in, encouraging attendance, and being present at an animal fight. The United Kingdom also prohibits supplying, publishing, showing or possessing with the intention of supplying a video recording of an animal fight.[58]

The United States federal government has also used indirect means of combating animal fighting ventures. The Animal Welfare Act prohibits use of the United States mail for promoting animal fighting ventures with the postal service making such materials non-mailable.[59] Organized cruelty, such as animal fighting, generally requires preparation, promotion and advertisement. Those engaging in animal fighting must obtain the animals, purchase paraphernalia used to train the animals or to "enhance" the fight, locate an arena for the fight, and solicit spectators to attend the fight. If avenues to advertise are cut off, animal fighting will diminish, as will animal suffering. Making such materials non-mailable can have a deterrent effect on animal fighting enterprises, although in this age of the internet and alternative means of communication it is perhaps less effective than it was previously. Nevertheless, it is useful to target secondary sources to better protect animals.

Another example of protecting animals indirectly from cruelty, similar to the United Kingdom provision, was 18 U.S.C. Section 48 that outlawed the knowing creation, sale, or possession of a "depiction of animal cruelty with the intention of placing that depiction in interstate or foreign commerce for commercial gain ... [unless the depiction] has serious religious, political, scientific, educational, journalistic, historical, or artistic value."[60] From the legislative history of the statute, the primary target of the law was so-called "crush videos," named for the horrifying depictions of women inflicting torture on animals while wearing high heeled shoes designed to appeal to fetishists who gain sexual gratification from watching such images. The cries from the animals, obviously in severe pain, can be heard on the videos.[61] However, the law was drafted more broadly, outlawing all depictions of animal cruelty that do not serve some redeeming purpose. Section 48 defined

depiction of animal cruelty [as] any visual or auditory depiction ... of conduct in which a living animal is intentionally maimed, mutilated, tortured, wounded, or killed, if such conduct is illegal under Federal law or the law of the State in which the creation, sale, or possession takes place, regardless of whether the maiming, mutilation, torture, wounding, or killing took place in the State.

Section 48 exempted depictions of animal cruelty that have "serious religious, political, scientific, educational, journalistic, historical, or artistic value" from prosecution. Thus, this law targeted those individuals who profit from the sale of such material by outlawing the creation, sale, and possession of these materials. Such laws are especially critical because the videos are integral to the illegal conduct and direct enforcement of the cruelty laws against those who inflict such torture is nearly impossible. Animal crush videos are filmed clandestinely and the individuals involved are never "seen." Moreover, the primary if not only purpose to torture these animals is to reap the profits from sales of these videos. Thus, these videos provide the economic incentive for, and are integrally related to, the underlying criminal acts.

This statute was quite controversial. Persons prosecuted under this law argued that prosecution infringes their constitutional right to freedom of expression, and eventually won. Difficult issues arise when criminal laws, designed to promote public health, safety and morals (for example, protecting animals from cruelty) arguably infringe on citizens' constitutional rights.

When anti-cruelty statutes arguably infringe defendants' constitutional rights

Infringing speech

Section 48 did not target the act of inflicting cruelty on an animal, but rather the creation, sale and possession of materials that depict such acts. As a result, defendants prosecuted under the law argued that this law infringed their right to free speech. In March 2004, Mr. Stevens was indicted under Section 48 for advertising and selling videos depicting animal fighting, "hunting excursions in which pit bulls were used to 'catch' wild boar, as well as footage of pit bulls being trained to perform the function of catching and subduing hogs or boars. This video include[d] a gruesome depiction of a pit bull attacking the lower jaw of a domestic farm pig."[62] In January 2005, Stevens was convicted and sentenced to 37 months in prison and three years of supervised release. He appealed on the ground that Section 48 infringed his right to freedom of speech. The Third Circuit, sitting en banc, agreed.

Two primary issues determined whether the statute was constitutional. First, does such speech deserve constitutional protection? Arguably, if it has no social value and itself is harmful, like obscenity or child pornography, it may not. Second, even if it does deserve protection, is the government's interest in protecting animals from cruelty compelling and the means used narrowly tailored to justify infringement on a defendant's free speech rights? The Third Circuit, en banc, with three judges in dissent, held the statute unconstitutional. In April 2010, the United States Supreme Court agreed, finding Section 48 "substantially overbroad, and therefore invalid under the First Amendment."[63]

Should depictions of animal cruelty be included in the very limited classes of speech deserving no protection under the constitution? On the one hand, free speech is the bedrock of American tradition, few human rights, if any, are deemed as important as the right to speak freely. As a society founded on free market principles, Americans strongly believe in the marketplace of ideas as the primary means by which truth and justice will prevail. When the government begins regulating what citizens may say, the marketplace is compromised and valid dissenting or minority views are lost. Moreover, if we allow the government to define unacceptable speech, how do we prevent the government's possible abuse of that power?

On the other hand, certain speech has no value and in fact does harm to society, and the courts have determined that certain very limited categories of speech with such characteristics are unprotected. If the conduct depicted in the materials is itself cruel and unlawful – acts of actionable animal cruelty – then outlawing the creation, sale and possession of the materials arguably does not present much of a threat to First Amendment principles. Moreover, if the materials are created and sold in order to debate the illegality of the underlying conduct, the materials arguably have either educational or political value and are protected under the statute. Thus, the marketplace of ideas would still flourish. Under this rationale Section 48 should survive.

The Supreme Court disagreed and refused to carve out an exemption to protected speech for depictions of animal cruelty as defined under the statute. The Court held that a "free-floating" test for First Amendment coverage based upon a simple balancing of the value of the speech against its societal costs is "startling and dangerous."[64] According to Chief Justice Roberts, although the Court has described historically unprotected categories of speech in this manner, the Court has never adopted such description as a definitional test. The Court held that the historical categories of unprotected speech that include obscenity, defamation, fraud, incitement and speech integral to criminal conduct are well-defined and narrowly limited and there is no evidence that "depictions of animal cruelty" is among them.

However, even if the speech is protected, how should the court weigh protecting animals from cruelty against a defendant's freedom of speech? Courts strictly scrutinize laws that infringe rights of speech but allow infringement if the government is seeking to promote a compelling purpose in a narrowly tailored manner pursuant to the least restrictive means available. The Third Circuit majority characterized the government's interest as "preventing cruelty to animals that state and federal statutes *directly* regulating animal cruelty under-enforce,"[65] and in turn held that this was not compelling when compared to the interest in preventing animals from cruelty more directly. It would appear that under such an analysis, *any* law that uses indirect methods to regulate otherwise illegal conduct would by definition not be classified as serving a compelling interest because they are "secondary" to the law that is targeting the illegal conduct directly. In a classic catch-22 style, the court stated that if the government interest is to protect animals from cruelty directly and is thus compelling, then the law is not narrowly tailored nor uses the least restrictive means available; for example, directly criminalizing the underlying conduct. The dissent, on the other hand, agreed with the government that the interest is to protect animals from intentional and wanton acts of physical harm which is compelling as demonstrated by the "overwhelming body of law across the nation aimed at eradicating animal abuse."[66] The dissent would have upheld the law as narrowly tailored.

The Supreme Court struck the statute as unconstitutional after determining that the language of the statute is of "alarming breadth."[67] According to the Court, the statute reaches conduct that the law does not deem cruel, including hunting and fishing or the humane slaughter of a stolen cow, because the definition of depictions of animal cruelty encompasses all acts that are illegal, whether pursuant to the anti-cruelty law or another law. The Court explained that many laws regarding animals are not designed to guard against animal cruelty but instead are to protect the loss of endangered species, the health of human beings, or to raise revenue and prevent accidents. Since the definition only requires that the conduct be illegal but not explicitly cruel, the statute cannot be upheld as constitutional. Moreover, the Court explained that the exemptions clause does not remedy the statute's overbreadth unless the Court were to agree to "an unrealistically broad reading"[68] of the clause, which it declined to do. The Court concluded by stating that they "do not decide whether a statute limited to crush videos or other depictions of extreme animal cruelty would be constitutional."[69]

Justice Alito was the lone dissenting voice of the Court. Interpreting the statute in light of legislative intent, Justice Alito explained that Section 48 "was enacted not to suppress speech, but to prevent horrific acts of animal cruelty"[70] and that the majority's "interpretation is seriously flawed."[71] Justice Alito found that the term "depictions of animal cruelty" covers conduct that is illegal under the anti-cruelty laws and not other laws unrelated to cruelty. However, even if hunting videos, for example, were covered, they would fall easily within the exemptions clause. Moreover, Justice Alito explained that the Court's holding in *Ferber*, which carved out child pornography from First Amendment protection, dictates a similar conclusion in *Stevens*. Violent crush videos and depictions of brutal animal fights should not be protected. The Justice stated that although "protecting children is unquestionably *more* important than protecting animals, the Government also has a *compelling* interest in preventing the torture depicted in crush videos."[72]

Justice Alito also found that depictions of brutal animal fights easily meet the *Ferber* test as well. He stated:

> First, such depictions ... record the actual commission of a crime involving deadly violence. ... Second, Congress had an ample basis for concluding that the crimes depicted in these videos cannot be effectively controlled without targeting the videos ... [and the] commercial trade in videos ... is an integral part of the production of such materials ... Finally, the harm caused by the underlying criminal acts greatly outweighs any trifling value that the depictions might be thought to possess. . . . For these dogs, unlike the animals killed in crush videos, the suffering lasts for years rather than minutes. As with crush videos, moreover, the statutory ban on commerce in dogfighting videos is also supported by *compelling* governmental interests in effectively enforcing the Nation's criminal laws and preventing criminals from profiting from their illegal activities.[73]

On the one hand, this decision seriously threatens the lives of helpless animals because no longer are these depictions outlawed, which allows the trade in these horrific videos to flourish. In 1999 prior to the statute, some 3,000 separately produced animal crush videos were found on the internet. Soon after the statute was enacted the market dried up.[74] Soon after the Third Circuit decision held the statute unconstitutional, the videos resurfaced, and with the Supreme Court ruling there is nothing to prevent their production and sale. On the other hand, the Court interpreted the statute so broadly that there is much room to narrow its construction and arguably enact a law that both outlaws depictions of animal cruelty (properly defined) and satisfies the First Amendment.

Soon after the *Stevens* decision was released, a bill aimed at prohibiting the sale and distribution of "animal crush videos" depicting senseless and vicious animal cruelty was introduced in the House of Representatives.[75] The Prevention of Interstate Commerce in Animal Crush Videos Act of 2010 begins by stating that the government has a compelling interest in preventing animal cruelty and that given the clandestine nature of certain acts of intentional cruelty that frustrates prosecution of the crime, this Act is necessary to protect animals from such heinous acts. The Act defines "animal crush video" as "any *obscene* photograph, motion-picture film, video recording, or electronic image that depicts actual conduct in which one or more living animals is *intentionally crushed, burned, drowned, suffocated, or impaled* in a manner that would violate a *criminal prohibition on cruelty to animals* under Federal law or the law of the State in which the depiction is created, sold, distributed, or offered for sale or distribution."[76] The italicized text notes the very narrow language of the Act drafted specifically to address the concerns of the Supreme Court. Moreover, in the findings Congress states that

> These criminal acts constitute an integral part of the production of and market for so-called crush videos and other depictions of animal cruelty. (5) The creation and sale of crush videos provide an economic incentive for, and are intrinsically related to, the underlying acts of the criminal conduct. (6) The United States has a long history of prohibiting the interstate sale of obscene and illegal materials. (7) Animal crush videos appeal to the prurient interest and are obscene.

This language is designed specifically to target speech that the Court has held unprotected, speech that is not only obscene but is integral to criminal conduct. Of course, Stevens would not have been prosecuted under this statute because his videos involving animal fighting are not outlawed. Whether the Court would uphold a broader statute that includes other depictions of acts that are criminalized in all 50 states as cruel is debatable. What is clear is that Congress will play it safe with this bill and attempt to eradicate the most egregious videos, and perhaps another day will target other videos, like those sold by Stevens.

As a brief aside, an interesting contrast to Section 48 is 18 U.S.C. Section 43 entitled the "Animal Enterprise Terrorism Act" (AETA) that prohibits the use of interstate commerce for the purpose of "interfering with the operations of an animal enterprise;" for example, any entity that uses animals for any purpose, "by a course of conduct involving *threats*, acts of vandalism, property damage, criminal trespass, *harassment or intimidation*."[77] This extremely broad statute arguably infringes the speech rights of advocates seeking to *protect* animals from abuse in a variety of settings, such as research institutions or factory farms.[78] Several persons prosecuted under the AETA have challenged the statute unsuccessfully. After *Stevens*, would this statute withstand the Court's scrutiny? Only time will tell.

Infringing religious practices

Prosecution under animal cruelty laws not only raises concerns of infringing defendants' speech but also of infringing their religious freedom. Religion plays an important role in how humans view animals. Religious beliefs for some dictate how much protection, if any, animals should receive and the moral or ethical justification for such protection.[79] In a diverse society, there are citizens of a variety of religious beliefs, and sometimes enforcement of the law can infringe upon a person's religious practices. This is especially true in the context of anti-cruelty laws. Specific religious practices, such as animal sacrifice or religious slaughter, may violate the anti-cruelty laws of the jurisdiction. How should the law address this conflict between the anti-cruelty laws and religious practices and beliefs?

In the United States, the Establishment and Free Exercise Clauses of the First Amendment are of paramount importance and enforce the principle of separation of church and state. The Establishment Clause "forbids an official purpose to disapprove of a particular religion or of religion in general."[80] Often these cases address government efforts to benefit or officially endorse a religion. The Free Exercise Clause protects citizens from discrimination or persecution because of their religious beliefs or practices.

The United States Supreme Court, in 1993, had the opportunity to address whether prosecution for the religious sacrifice of animals under the anti-cruelty law violated the defendant's freedom of religion in the case of *Church of the Lukumi Babalu Aye* v. *City of Hialeah*.[81] Santeria adherents had been persecuted in Cuba and many moved to Florida to flee such persecution.

> The Santeria faith teaches that every individual has a destiny from God, a destiny fulfilled with the aid and energy of the *orishas* ... [who] are powerful but not immortal ... [and] depend for survival on the sacrifice ... [of animals including] chickens, pigeons, doves, ducks, guinea pigs, goats, sheep, and turtles. The animals are killed by the cutting of the carotid arteries in the neck. The sacrificed animal is cooked and eaten, except after healing and death rituals.[82]

Many Floridians in the city of Hialeah found these sacrificial practices obnoxious and soon a law was proposed to outlaw animal sacrifice. Municipalities in Florida are prohibited from enacting local laws that conflict with state law. The state anti-cruelty law already outlawed the unnecessary or cruel killing of an animal. The attorney general advised the city that animal sacrifice was against the state law, given animal sacrifice is arguably "unnecessary" as defined as "done without any useful motive, in a spirit of wanton cruelty or for the mere pleasure of destruction without being in any sense beneficial or useful to the person killing the animal."[83] Thus, Hialeah could outlaw animal sacrifice without conflicting with state law.

The city enacted a law that stated: "No person shall own, keep or otherwise possess, sacrifice, or slaughter any sheep, goat, pig, cow or the young of any such species, poultry, rabbit, dog, cat, or any other animal, intending to use such animal for food purposes." The law exempted any licensed establishments that slaughtered for food purposes any animals raised for food purposes where properly zoned. The law also prohibited any person from sacrificing any animal within the city limits and prohibited the slaughter of animals outside of areas zoned for slaughterhouse use unless for the sale of "small numbers of hogs and/or cattle per week in accordance with an exemption provided by state law." The stated justification for the law was that the sacrifice, or slaughter of animals outside of properly zoned, areas "was contrary to the public health, safety, welfare and morals of the community."[84]

Under the free exercise doctrine as defined in *Employment Div., Dept. Human Res. of Or.* v. *Smith*,[85] neutral laws of general applicability "need not be justified by a compelling government interest even if the law has the incidental effect of burdening a particular religious practice."[86] However, if the law is not neutral or of general applicability; for example, the law is targeting a religion or religious practice, the government must prove that the law is justified to satisfy a compelling governmental interest and be narrowly tailored to advance that interest.

The court held that the Florida law was *not* neutral and of general applicability but rather was gerrymandered in order to target the Santeria practices. Moreover, the law was both over-inclusive and under-inclusive in its approach to advancing the legitimate interests of protecting the public health and preventing animal cruelty. The court explained the law is over-inclusive because the protection of animals from cruelty could be achieved by regulating the "conditions and treatment, regardless of why an animal is kept"[87] rather than prohibiting sacrifice per se. Moreover, if the method used to kill the animal causes the cruelty, the law should regulate the method of, not the purpose behind, the slaughter. Furthermore, the law was considerably under-inclusive because it allowed the killing of animals for many other reasons, including fishing, hunting, extermination of pests, and in the interests of medical science, all of which may result in pain and suffering to the animal. Interestingly, the court rested its decision on the basis that the law was not narrowly tailored while being careful to describe the nature of the government's interest in protecting animals from cruelty as legitimate rather than compelling.

An interesting question was raised by Justice Blackmun.[88] He asked, would the constitution require that the state exempt the religious sacrifice from the general anti-cruelty law? Presumably, if the anti-cruelty law is neutral and of general applicability, it may incidentally burden a religious practice without infringing one's free exercise of religion. However, the court is split on the *Smith* rule's vitality as precedent. Justice Scalia, a strong believer in the notion of *stare decisis*, nevertheless stated that the *Smith* rule "may be reexamined consistently with principles of *stare decisis*,"[89] and presumably would denounce the rule. Moreover, even if the court required that the government interest be compelling in order to incidentally burden religious practice, would or should animal cruelty satisfy that standard?

Compare this U.S. case with one from the European Court of Human Rights (ECHR). In *Cha'are Shalom Ve Tsedek* v. *France*,[90] the ECHR, with seven judges dissenting, held an applicant's religious rights were not violated when denied an exemption under the French humane slaughter laws for their religious slaughter.[91] The Jewish liturgical association Cha'are Shalom Ve Tsedek, which represents a minority group within the Jewish community, was denied an exemption to perform religious slaughter in France. The association was founded in 1986 when it split from the Jewish Consistorial Association of Paris (ACIP) to practice a strict orthodoxy. Their religious slaughter practices not only require that the animals not be stunned prior to slaughter, in violation of the slaughter laws, but also that the lungs of the animal be examined, a procedure not included under the ACIP religious slaughter. Meat slaughtered in this manner is called *glatt*. The lower courts had refused to recognize the group as an approved religious association and noted that other Jewish groups had been granted exemptions to practice the religious slaughter. The applicant association claimed that the denial violated their religious rights as protected in Article 9 of the European Convention on Human Rights.[92]

The ECHR majority held that religious slaughter is covered under Article 9 and noted that the French law authorizes religious slaughter as generally practiced by the Jewish faith. However, permitting only religious slaughterers authorized by approved religious bodies does not interfere with religious practices. The court stated that France has an interest to avoid unregulated slaughter and to require that it be performed in slaughterhouses supervised by the public authorities. In sum, since other Jewish religious bodies had been granted an exemption for religious slaughter, and since *glatt* meat may be imported from Belgium and was available from a few ACIP butchers, the association could obtain *glatt* meat and therefore did not have their religious rights under the Convention violated. The dissenting judges argued that pluralism, even within religious groups, must be respected and that the association had a right to perform their religion through their own religious slaughter.

The case of religious slaughter arguably presents a slightly different balancing than that of religious sacrifice. For example, in the case of the Santeria faith, the religion mandates the sacrifice of animals. However, in the case of the Jewish faith, the religion does not require that Jews eat meat. However, if an observant Jew chooses to eat meat, the animal must be slaughtered pursuant to the proper methods. When balancing the interests of the religious observers against that of the government, does or should such a distinction make a difference? In the United States, the First Amendment prohibits the government from substantially burdening the exercise of religion. The religious practice at issue need not be a mandatory religious practice it need only be a central or important practice. Thus, laws that prohibit the method of slaughter or the method of sacrifice would implicate the First Amendment and would be deemed to burden the right of religious practice. However, whether the law burdens a mandatory religious practice or only an important religious practice should be relevant to the balancing of the burden against the government's purpose. Arguably, a law that burdens an important religious practice should be sustained under a lesser showing of governmental purpose than one that burdens a mandatory religious practice.

A federal case from New York, *Jones* v. *Butz*,[93] presents an interesting twist on these cases. "Helen Jones as next friend and guardian for all livestock animals now and hereafter awaiting slaughter in the US" and others brought suit in federal district court against the Secretary of Agriculture, claiming the federal Humane Method of Slaughter Act (HMSA) religious slaughter provisions violated the First Amendment Establishment and Free Exercise Clauses. The plaintiffs, as taxpayers, professed commitment to two principles: the humane treatment of animals and the separation of church and state. They claimed that the religious slaughter provisions of the HMSA harmed them because they were not able to distinguish meat products slaughtered pursuant to the exemption from other meat products. As a result, they had to abstain from eating meat or had unwittingly eaten meat from animals slaughtered pursuant to the exemption. The court held that their commitment was deeply held and sincere and that their allegations of injury as consumers and citizens to their "moral principles and aesthetic sensibilities"[94] were sufficient to confer standing.

The court then turned to the merits of the case and upheld the HMSA. The plaintiffs argued that by failing to require that the animal be rendered insensible to pain before they are "shackled and hoisted, the statute's provisions permitting ritual slaughter are offensive to and inconsistent with the humane purposes of the Act and have a special religious purpose,"[95] violating the First Amendment. The court held that Section 2(a), defining methods with prior stunning, and (b), defining religious slaughter without prior stunning, are alternative methods of humane slaughter. Section 2(b) is not an exception to humane slaughter. Thus, Congress was not deferring to this method out of deference to religious practice per se but rather because they found it an alternative humane method. In fact, the religious slaughter method of a quick, deep stroke across the throat with a perfectly sharp blade has been widely recognized as a humane method of slaughter.[96]

The plaintiffs' claim, however, concerns the shackling and hoisting of the animals prior to stunning. United States Department of Agriculture (USDA) regulations prohibit putting an animal down while on the ground for human health reasons. Thus, the animal is shackled and hoisted by its legs and turned upside down in order to administer the throat cut. Ironically, shackling and hoisting is *not* part of the Jewish ritual practice. "In Israel, and indeed, in the old traditional Jewish method, the animal would be laying on its side, and the throat would be cut on the floor."[97] Moreover, the Committee on Jewish Law & Standards of the Rabbinical Assembly has held that shackling and hoisting is a *violation* of Jewish laws forbidding cruelty to animals.[98] Thus, Congress' determination that the practice is humane was based on incomplete information. Arguably, if Congress had considered the prior shackling and hoisting it would not have found it to be an alternative humane slaughter method. That apparently would not have altered the decision in the case as the court further held that Congress could have allowed the method out of deference for religious practice,[99] but the inhumane treatment is *not part* of the religious practice. Today there are humane restraint systems that allow the animal to be placed in a restrainer that holds it in a comfortable, upright position for the procedure.[100] Since Congress apparently allowed the practice because they thought it was humane, not accounting for the shackling and hoisting, and the shackling and hoisting violates Jewish law, it would seem that Congress should require under the HMSA that religious slaughter methods use these restraint devices to prevent needless suffering and comport with religious principles.[101]

Finally, it is interesting to compare these cases where anti-cruelty and welfare laws allegedly infringe constitutional rights, with cases where the conservation laws, such as the Bald and Golden Eagle Act (BGEA), allegedly infringe constitutional rights. The BGEA prohibits the taking of any bald or golden eagle except for limited purposes, including for the religious purposes of Indian Tribes as eagle feathers are an integral part of Native American religions. The Fish and Wildlife Service will grant an application for a religious taking permit only when it "determine[s] that the taking ... is compatible with the preservation of the bald and golden eagle."[102] There are two listed criteria for this determination – the likely "direct or indirect effect" of the permit on the eagle population, and whether the applicant is a Native American with a "bona fide" religious use. Native Americans have argued that the permit requirement to seek an exemption from the BGEA is an unreasonable burden on their free exercise of religion. Note that the HMSA expressly allows for religious slaughter, thus butchers are not required to seek an exemption under the HMSA. Here, the BGEA requires that Native Americans expressly seek an exemption before taking the birds.

The courts have upheld the BGEA as being the least restrictive means for serving a compelling government interest. What is particularly interesting is the court's explanation of the nature of the government's interest in protecting the bald and golden eagles. Even after the eagles were removed from the list of endangered or threatened species in 2007, the Ninth Circuit held that the government retained a *compelling* interest because Congress had explained that "the bald eagle is [not] a mere bird of biological interest but a symbol of the American ideals of freedom."[103] Moreover, the "bald eagle would remain our national symbol whether there were 100 eagles or 100,000 eagles. The government's interest in preserving the species remains compelling in either situation."[104] Since the law is protecting an animal for the direct benefit of humans, because the bird is the national symbol (in contrast to protecting animals from cruelty) the interest is compelling. In fact, wildlife generally is *not* protected under the anti-cruelty laws, as we will soon see.

A case study: cruelty to animals used for food – comparing the New Jersey and Israeli approaches

The case of animals used for food presents an interesting analysis of how humans view cruelty to animals. In the United States, animals used for food generally are exempt from the cruelty laws and their treatment not regulated by law. The result is that most animals used for food are subject to severe cruelty every day of their lives.

New Jersey is the first and only state to date to attempt to system-atically address institutionalized cruelty to animals used for food. In 1996, the New Jersey legislature instructed the New Jersey Department of Agriculture (NJ DOA) to implement regulations establishing standards for the humane care, treatment, raising, keeping, marketing, and sale of domestic livestock within six months.[105] Regulations were finally adopted eight years later in 2004. Soon thereafter, the New Jersey Society for the Prevention of Cruelty to Animals (NJ SPCA) filed suit against the NJ DOA challenging the regulations as arbitrary and unreasonable. The decision of the court reads like a chapter out of the book *Alice in Wonderland*. The court upheld as "consistent with the agency's legislative mandate, and [as] neither arbitrary, nor unreasonable,"[106] the "humane" regulations codifying "farming practices that the record overwhelmingly demonstrates cause severe hunger, pain, stress, disease, and even mortality in animals."[107] These "humane" practices include tethering 450-pound calves in individual crates a mere 26 inches wide; sow gestation crates that prevent pigs from turning around, much less performing basic, natural movements; mutilation practices of tail-docking, castration, debeaking, and toe-trimming, without the use of anesthesia; the forced feeding of ducks and geese for foie gras, and all "routine husbandry practices."[108] Recognizing that the plaintiffs had ample support in the literature and in the veterinary community that these practices are indeed "inhumane," the court nevertheless found that the record presented a "divergence of opinion" and that deference to the agency's expertise and experience was appropriate. The New Jersey Supreme Court ultimately affirmed in part and reversed in part this decision.[109]

The decision demonstrates a complete lack of respect and compassion for animals used for food by the agency tasked with protecting the animals and by the court tasked with enforcing the law. The New Jersey court did not have the courage to question, much less strike down, the agency's codification as "humane" numerous inhumane practices that promote the economic viability of the farming industry. That decision stands in stark contrast to the Israeli Supreme Court decision that three years earlier had outlawed as criminally cruel the force-feeding of ducks and geese used to produce foie gras in that country.[110] The Israeli high court was willing to disregard "accepted animal husbandry practices" and protect ducks and geese from the suffering imposed by force-feeding at the expense of eradicating an entire industry.[111]

Admittedly there is much to distinguish the two decisions apart from the specific legal question presented. The Israeli legal system, including the substantive laws and the role of the courts, is vastly different from that of the United States. Moreover, the cultural and religious beliefs of the two countries are distinctly different. Thus, different decisions, rationalizations, and outcomes are expected. Nevertheless, it is hard to reconcile the widely divergent notions of what constitutes the "humane" versus the "cruel" treatment of sentient beings regardless of one's legal, cultural, or religious beliefs. The following discussion compares and contrasts the two decisions focusing on the courts' general attitudes toward animals and their treatment.

The animals – who they are and our obligations to them

Perhaps the most striking difference between the two opinions is their discussion of the animals who are the subject of the laws. The New Jersey decision, joined by all three judges of the panel, is exceedingly legalistic in its tenor and never once mentions the nature of animals as sentient beings. Throughout the opinion the court cites exclusively to statutes, regulations, cases, and the organizations whose standards the Department relies upon in establishing the regulations, including the American *Veal* Association's Guide, the National *Pork* Board, and the California *Pork* Industry Group. The mere names of these organizations demonstrate their conception of the animal as a meat product to be sold rather than a sentient creature to be protected.

In contrast, the Israeli opinion of Justice T. Strasberg-Cohen, joined by Justice E. Rivlin,[112] identifies the animals and recites the obligation, rooted in Jewish law and part of the Israeli legal system, to protect them by prohibiting cruelty to them. Most strikingly, the Justice explains the basis of this obligation and explores the range of philosophical views on man's relation to animals, from Descartes' belief that "man is the master of all and has the right to subject animals to his will" to Tom Regan's and Peter Singer's beliefs that animals have interests of their own, citing their works laying the moral and ethical foundation for considering the interests of animals to be free from pain and suffering.[113]

When the opinion turns to the more strictly legal question presented, Justice Strasberg-Cohen notes the system's checks and balances between the Minister of Agriculture's responsibility to issue regulations taking into account the agricultural needs of society with the Knesset's Education and Cultural Committee's responsibility to ensure that proper weight is given to the interest of animal protection when enacting the regulations. Note that there is no separate entity in New Jersey tasked with protecting the animals' interests. The New Jersey Department of Agriculture is responsible for setting the standards for the humane regulation of animals used for food although its goal is to promote the agricultural industry.

The court's role

The New Jersey court defines its role in light of its standard of review.[114] It is clear from the outset that the regulations will stand as the court summarizes the "well established" policy that "agency regulations are presumed valid" and reasonable, that the court's review is "severely limited" and should "intervene only in the rare circumstances in which an agency is clearly inconsistent with its statutory mission or other state policy."[115] Furthermore, "administrative rulemaking does not require findings of fact sufficient to justify the regulations . . . Instead, 'facts sufficient to justify the regulation must be presumed'"[116] and the burden lies with the party challenging the regulations to establish that the requisite facts do not exist. Finally, the court is to "accord substantial deference to the interpretation an agency gives to a statute that it is charged with enforcing because 'agencies have the specialized expertise necessary to enact regulations dealing with technical matters and are particularly well equipped to read and understand the massive documents and to evaluate the factual and technical issues that . . . rulemaking would invite.'"[117] In fact, "findings may be based on an agency's expertise, without supporting evidence" and are "especially important to the judicial review process when the court does not share the agency's expertise."[118]

In contrast, the Israeli Justice states that the "Court's intervention concerning regulations that were approved by a Knesset committee will be done only rarely. Yet, it must be noted that regulations are not immune to judicial review, even if they received parliamentary sanction. Regulations can be annulled when they deviate significantly from the purpose of the law."[119] Ultimately, it is the Court's duty to "carefully examine all the facts . . . consider the opinions of experts in several fields, the legal situation in various countries and in the international community, the domestic legal situation, and the extra-legal questions raised . . . [and determine if] the regulations deviate significantly from the purpose of the law."[120]

The stage is diametrically set in the two cases. In the New Jersey opinion the reader is never introduced to the animals who are the subjects of the court's opinion, much less their interests, and the bar is set extraordinarily high for the challenger. In the Israeli opinion, the reader is introduced to the animals and their interests in their own right and, while the bar is high, the court makes clear that it is willing to do justice.

"Accepted agricultural practices"

Another stark contrast between the two opinions is the treatment of "accepted agricultural practices" under the law. The Israeli court was clear, stating "[l]ong accepted agricultural practices do not have immunity from the law [against abuse]."[121] In contrast, the New Jersey regulations approve as humane "routine husbandry practices" defined as "those techniques commonly taught by veterinary schools, land grant colleges, and agricultural extension agents for the benefit of animals, *the livestock industry, animal handlers and the public health* and which are employed to raise, keep, care, treat, market and transport livestock."[122] The challengers argued that this creates a vast exemption for such practices regardless of whether they are humane. In response, the NJ DOA explained that "[t]hese standards are not intended to modify those routine animal agricultural practices that are performed each day by farmers in New Jersey, but rather to protect animals from only those practices that are inhumane or cruel."[123] This suggests that most, if not all, routine animal agricultural practices are considered humane by the NJ DOA because they are, in fact, routine. Notably, most state cruelty laws exempt "accepted animal husbandry practices,"[124] implying that the practices would likely violate the state's cruelty laws. If not, there would be no need for an express exemption. Here, the NJ DOA is not exempting these practices from the state's cruelty laws but instead endorsing them as humane.

Humane versus cruel

The fourth striking contrast between the two decisions is the definition of what is criminally outlawed as cruel versus what is endorsed as humane. Note that these terms – "cruel" and "humane" – do not represent two discrete categories but rather denote points along a continuum. The New Jersey regulations were designed to set standards for the humane treatment of the animals and the court was to decide if they were consistent with attaining that goal. The regulations defined the term "humane" as a practice "marked by compassion, sympathy and consideration for the welfare of animals," but expressly stated that the regulations established "the *minimum* level of care that can be considered to be humane . . . not best management practices ... [T]he standards are intended to serve as the *baseline* for determining inhumane treatment."[125] Thus, the regulators expressly chose to set standards that just barely may be considered "humane" to "[ensure] that any ... treatment that falls below these standards can be accurately identified and swiftly addressed."[126] The regulations do not explain why higher standards would create problems of inaccurate identification or slower enforcement. One reason might be that treatment falling below a higher standard would not be understood as inhumane by the average farmer. However, ignorance of the law is never an excuse to avoid the law and sufficiently specific standards would provide adequate notice of what is considered humane under the law. A second reason might be that the authorities would be overwhelmed by the number of violations if a higher standard is set, making it hard for them to adequately address them all. This suggests not only that the current accepted practices do not treat animals humanely, but also that law enforcement only has sufficient resources to correct the most severe inhumane practices. However, if minimum protection is all that the legislature intended, they could have demanded better enforcement of cruelty laws. Instead, the legislature demanded regulations that would require the humane treatment of animals, not merely the non-cruel treatment of animals.

In contrast, the issue before the Israeli court was whether the regulations, designed to "prevent the suffering of geese" during the production of foie gras, violated the criminal law prohibiting animal abuse. Since the Israeli animal cruelty law imposes a criminal sanction on the violator, the definition of criminal abuse should be lower than the definition of humane treatment at issue in the New Jersey case. Thus, the line delineating humane treatment in the New Jersey case, even if designed to set minimal standards, should be equal to or higher than the line delineating criminal abuse by the Israeli Court. This, however, was not the result.

In the New Jersey case, the force-feeding of geese was only one of the challenged and allegedly cruel practices permitted under the agency regulations. The court's analysis of the foie gras issue consumed *three paragraphs* of the court's 18-page opinion.[127] While the challengers argued that force-feeding geese for foie gras "unquestionably causes birds significant pain and suffering"[128] and thus is inhumane, the Department relied on the American Veterinary Medical Association (AVMA) House of Delegates' decision declining to prohibit the practice after finding "there was insufficient peer-reviewed evidence related to animal welfare during the production of foie gras."[129] The court held that "because there is no conclusive evidence that the practice is detrimental to animal welfare, the agency's decision not to ban it cannot be considered arbitrary or unreasonable."[130] Thus, the practice of force-feeding was not found inhumane.

Although the court spent very little time on the foie gras force-feeding procedure, they did perform a slightly more comprehensive analysis of the effect of other alleged cruel practices allowed under the humane standards. These offer additional insight into the court's interpretation of humane. The New Jersey court used a balancing test – balancing the alleged advantages and disadvantages of the practices – to determine if they were reasonable and thus within the agency's discretion to define what is humane. Taking sow gestation crates and tethering of sows as an example, the government explained that on balance, housing sows in crates and tethering them is better than housing sows in traditional group pens where many animals are placed together and, although free, have very little space, if any, to move about. The advantages are that "individual stalls ensure that fighting between sows is minimized; provide each sow with a full ration of food; and permit swine producers to identify signs of morbidity, and treat the sow accordingly." The agency relied on studies that indicated there are fewer injuries to sows individually housed since group pens tend to result in increased aggression and biting. The agency ultimately determined that the "[r]eproductive performance is in general better in stall-housed versus group-penned sows ... In addition, the health of the animals is more easily monitored when housed individually."[131]

The challengers explained that many disadvantages accompany the tethering and use of sow gestation crates. These include: preventing pigs from performing the most basic, natural movements, such as simply turning around, and causing "physical and psychological ailments, such as decreased muscle mass, decreased bone strength, cardiovascular disease, boredom, apathy and stereotypic behavior."[132] Further, they noted that several state and foreign governments have denounced gestation crates as inhumane. In response, the government claimed that such effects have not necessarily been conclusively proven when compared to keeping the animals in traditional group pens. The court agreed with the government.

The NJ SPCA also challenged the accepted "mutilation practices" of tail-docking, debeaking chickens and turkeys, toe-trimming of turkeys, and castration of pigs without anesthesia as humane under the regulations.[133] The challengers noted that each mutilation practice is for the benefit of the human owner, not the animal. For example, tail-docking allegedly improves "cow hygiene by reducing the incidence of udder infection" and maintains "milk quality by reducing ... fecal contamination."[134] However, the practice causes pain and prevents the cows' ability to perform natural behaviors. Similarly, castration without anesthesia is "devastatingly painful"[135] and does not benefit animal welfare. Its purpose is to avoid the "tainting of pork with foul odors and off flavors" and to reduce "aggressiveness and handling problems associated with intact males;" both are purely economic reasons that benefit the animal handlers, not the animals.[136] Similarly, the debeaking and toe-trimming of birds are extremely painful to the bird but are allegedly needed to protect the flock from cannibalism and other types of bird behaviors that are caused by the overcrowding of the birds into a very small space.[137]

The court nevertheless found that

> the agency is entitled to determine that ease of management and avoiding "off flavors" are consistent with the Legislature's expressed mandate directing the Department to develop standards governing "marketing and sale of livestock." That interpretation of its mandate is an exercise of agency discretion protected by law, and we decline to substitute our judgment on the relative benefits of castration for that of the agency.[138]

The court reasoned that since the regulations provide that such mutilations are to be "performed in a way that minimizes pain," they meet the humane standard.[139]

Did the agency and the court properly account for the animals' interests and serve the legislative mandate? First, the agency compared two arguably inhumane practices – crates and tethering with traditional group pens. The NJ Supreme Court on appeal did recognize that the controversy over the use of these mutilation practices "stems from the larger question of whether farm animals are to be raised in close quarters or in spacious and relatively unconfined surroundings."[140] The court referred to that question as a "philosophical debate about how farm animals are raised and kept in general,"[141] which in turn "cannot help but affect one's views about whether some of these procedures are, on the one hand, pointless and cruel, or on the other hand, necessary techniques for managing the livestock in one's care."[142] The court, however, found that this debate "is not addressed in the statute," but rather "the agency was charged with finding an appropriate balance between the interests of animal welfare advocates and the need to foster and encourage agriculture in this state."[143] This position contributes to the circular reasoning on which the NJ DOA relies. If one assumes that the legislature did not desire to alter the keeping of livestock in "close quarters" when it enacted a statute requiring the NJ DOA to implement humane standards of keeping the animals, then the mutilation practices to prevent the animals housed in this manner from harming each other are by definition necessary, which in turn makes them humane under the balancing test. This is a rather ironic conclusion.

Moreover, in balancing the advantages and disadvantages, human interests trumped. The advantages primarily serve to benefit the human handler and owner, by facilitating the feeding and the monitoring of the animals, avoiding tainting the meat, and increasing profit from enhanced productivity. The disadvantages all accrue to the animals and are severe, both physically and psychologically. Interestingly, in the discussion of the tethering of calves, the government noted: "tethered calves, when released into a large enclosure, actually perform more locomotive behavior than group housed calves – a strong indication that there is no delay in their muscular development."[144] Is it any wonder that calves who are continuously tethered with no ability to even turn around "perform more locomotive behavior" than the group-housed calves when released into a large enclosure? Rather than suggest there is thus "no delay in their muscular development," it may suggest that the calves are truly in desperate need of the ability to move at all.

In contrast, the Israeli Court devoted over *50 pages* to the foie gras issue alone. The Israeli legal system also uses a balancing test – the utilitarian test of purpose and proportionality – to define "abuse" by balancing "the degree of suffering caused to the animal, the purpose of this suffering, and the means used to achieve the purpose."[145] Under the Israeli purpose and proportionality test, the court weighed "the relevant 'agricultural needs' [which includes the interests of the farmers that produce foie gras] ... against the suffering inflicted on the animal, as well as the type of suffering and its severity."[146] The court found "no real disagreement that the practice of force-feeding causes the geese suffering."[147] The court described in detail the force-feeding procedure and considered the 1998 Report of the European Council's Scientific Committee concerning animal health and welfare that extensively examined the effect of force-feeding on the welfare of geese.[148] The report concluded that "force-feeding, as currently practiced, is detrimental to the welfare of the birds."[149] The court also canvassed other countries' policies with regards to the practice and found that France, Spain, and Belgium continue the practice, while Norway, Germany, Austria, Denmark, the Czech Republic, Poland, and Luxembourg all prohibit force-feeding. Moreover, while the Standing Committee of the European Convention for the Protection of Animals Kept for Farming Purposes in 1999 did not recommend ending the practice in countries where it already existed, they recommended that additional research on the effects of the process on the animals' welfare and on alternative methods for production be encouraged and that the practice should continue *solely* in countries where it was in existence prior to 1999.[150] The Israeli court did not consider the AVMA or any other American source after noting that most states in the United States (as well as Canada) generally "exclude ... accepted agricultural practices from the application of animal protection laws."[151] The court concluded that sufficient evidence proved that force-feeding is detrimental to the birds.

Moreover, although the New Jersey decision ended its analysis when it found no conclusive evidence that force-feeding of geese is detrimental to their welfare, the Israeli court continued its balancing test to determine whether the suffering caused to the geese was criminal abuse. The court held that while the regulations attempted to balance the interest of animal protection against agricultural needs by freezing the industry and establishing substantive standards regarding the process, they failed.[152] The court explained that while the purpose, the production of food, has greater weight than entertainment, for example, foie gras is a luxury foodstuff and certainly not necessary for human existence.[153] Moreover, the harm caused to the geese is insufficiently minimized by the standards imposed. Thus, the price paid by the geese to produce a delicacy is too high. The court ultimately held that the regulations be annulled, but suspended the application for 18 months to consider developments in the field and an appropriate policy regarding force-feeding geese.[154] The court ended by stating that if the foie gras industry is to continue, then regulations assuring the use of means that will significantly reduce the suffering of geese must be enacted. Today, foie gras production is banned in Israel.[155] This was a huge victory for the geese and a major step for Israel, since Israel had been among the top four producers of foie gras in the world.

Lessons learned

In sum, this comparison suggests several issues that must be considered if animals used for food are to be protected from cruelty. First, in the United States, the Department of Agriculture, an agency with the primary goal of promoting the agricultural industry not animal welfare, is responsible for enacting and enforcing humane regulations. In contrast, Israel has a system of checks and balances with separate agencies tasked to promote each goal: agricultural needs and animal welfare. Without a separate entity whose purpose is to promote animal welfare, the interests of the animals will not adequately be protected.

Second, the courts in the United States play an extremely limited role in reviewing challenges of agency regulations and are unwilling to do justice when an agency fails to adequately pursue their legislative mandate. In contrast, Israeli courts have more power to review agency decisions and a willingness to do justice. Third, the agency does an injustice to the animals when it relies on the standards of the agricultural industry itself (for example, the National *Pork* Board). It seems odd to defer to the views and practices of the regulated industry when establishing regulations to govern that same industry, especially if the regulations are by definition designed to protect the animals at some expense to the industries' financial bottom line. The welfare of animals used for food is so poor because the industry engages in practices that are cruel in order to make more profit. Thus, humane regulations must target the existing, accepted, cruel practices and not use them as a guide to define what is humane.

Fourth, in determining whether an alleged procedure detrimentally affects animal welfare (for example, tethering), the baseline should be conditions that are humane and promote the animals' welfare not an alternative inhumane procedure (for example, confined pens). Fifth, in determining whether a practice is humane the balancing of the advantages and disadvantages of the procedure should adequately reflect the animals' interests, not those of the handlers and industry. And finally, if the goal is to establish standards of humane care, the standards should promote the positive physical and emotional welfare of the animal not merely avoid the infliction of cruelty upon the animal.

What about wildlife?

Introduction

The category of wild animals is broadly defined. A federal conservation statute defines "fish or wildlife" to mean "any wild animal, ... including without limitation any wild mammal, bird, reptile, amphibian, fish, mollusk, crustacean, arthropod, coelenterate, or other invertebrate, whether or not bred, hatched, or born in captivity, and includes any part, product, egg, or offspring thereof."[156] State wildlife laws similarly have broad definitions such as "all species of invertebrates, fish, amphibians, reptiles, birds and mammals which are ferae naturae or wild by nature."[157]

With respect to animals in the wild, one American commentator has stated:

Not surprisingly, state laws relating to the humane treatment of wildlife, including deer, elk, and waterfowl, are virtually non-existent. This is primarily due to the fact that humans interact far less with wildlife than with domesticated species. Wildlife are not used for companionship, like pets, but for food and sport.[158]

This apparently is an accurate assessment of the laws protecting animals worldwide, as a second commentator from Australasia has noted that wild animal welfare law is sparse, bordering on the non-existent at the international level and moving down to the national level it is clear that the welfare of animals living in the wild usually receives far less attention than the welfare of [other categories of animals]. Often this area of the law derives unobtrusively, incidentally, or even accidentally from measures designed to conserve species.[159]

Anti-cruelty versus game laws

In the United States, the majority of states exempt wildlife from their cruelty laws and defer to state game laws for guidance in treating wildlife.[160] The cruelty law may either expressly exempt the entire class of wildlife, or exempt certain practices, such as hunting, fishing or trapping.[161] While it may seem that the two approaches are similar, their effects and underlying purposes are quite different. A cruelty law that exempts all wildlife allows humans to cruelly treat any wild animal in any manner and for any purpose. However, a cruelty law that exempts specific accepted practices at least allows for prosecution of cruelty to wildlife when the actor is not involved in the exempted activity. Thus, laws that exempt specific practices rather than entire classes of animals are generally more protective of animal interests. Further, one may argue that laws that exempt certain practices are designed to allow humans to engage in practices that serve important societal goals rather than intimating that certain classes of animals are not worthy of protection based solely on their identity. Most state game laws, however, do not independently outlaw the cruel treatment or killing of wildlife, thus leaving wildlife virtually unprotected from cruelty. Although, it must be noted that game laws often prescribe the methods of hunting and trapping permitted under the laws and have begun to outlaw certain methods that are deemed excessively cruel.

The relationship between the state game law and the state anti-cruelty law can be confusing and controversial if the anti-cruelty law does not expressly exempt wildlife or certain practices. For example, if there is no express exemption, does the anti-cruelty statute automatically govern wildlife? If so, what if the game laws prescribe conduct that otherwise would be deemed cruel under the cruelty statute? Do the game laws preempt the cruelty law or must the game laws come into compliance with the cruelty law? The following cases have addressed these and similar issues.

In *New Mexico* v. *Cleve*, the defendant was a rancher who "shot at least thirteen deer, five in the abdomen, and snared two others. In one of the snares, a fawn was caught by the neck and died of strangulation, probably within about five minutes of being caught. In the other snare, a spike buck was caught by its antlers and died of stress-related fatigue, starvation, or dehydration."[162] Cleve was charged with, inter alia, seven counts of animal cruelty, and convicted of two counts for the snared deer. Cleve argued that (1) the animal cruelty statute does not apply to wildlife, and (2) the game statutes pre-empt the animal cruelty law. The trial and appellate courts disagreed on both grounds. The appellate court explained that the game and cruelty laws exist

> for "different purposes". It is possible to illegally hunt game animals, but not to have been cruel in killing them ... Conversely, one could be convicted of cruelty to animals, but not of unlawful hunting of game animals ... [B]oth the cruelty statute and the game and fish laws and regulations are necessary to fully protect wild animals, and these two statutes can coexist ... [Moreover, the] game animals must be included within [the cruelty law] because to hold otherwise would leav[e] many animals unprotected [and] would create an unjust or absurd result.[163]

The New Mexico Supreme Court reversed the appellate court on both grounds, finding that (1) even though the cruelty statute refers to "all animals," it applies only to "domesticated animals and wild animals previously reduced to captivity,"[164] and (2) if it did apply to wild animals, the game laws would pre-empt the cruelty laws. Here is the court's explanation. Is it persuasive?

> The New Mexico cruelty statute defines cruelty as:
> A. torturing, tormenting, depriving of necessary sustenance, cruelly beating, mutilating, cruelly killing or overdriving *any animal*;
> B. unnecessarily failing to provide *any animal* with proper food or drink; or
> C. cruelly driving or working *any animal* when such animal is unfit for labor.[165]

Although the statute clearly applies to "any animal," the court, relying on a previous case that held that cock-fighting was not covered under the cruelty statute, held that the intent (if not the express language) was for the statute to apply only to brute creatures and work animals. The court based this finding upon a review of the statute as a whole, comparing other states' cruelty statutes, and the history of animal cruelty law in the United States. The court stated that "[c]learly, the Legislature did not intend to create a duty on the part of the public to provide sustenance to wild animals," as described in subsection B. Further, other similar statutes[166] within the same article and enacted at the same time refer to domestic and work animals, although they too refer to "any animal." Thus, the original intent of the legislature was to cover only domestic and captive animals under the cruelty statute.

The court then turned to the pre-emption issue since Cleve was charged and convicted under both the game and anti-cruelty laws. The court held that the legislature intended to create separately punishable offenses under the game laws and anti-cruelty laws because the elements of the two sets of laws differ and the purposes served are distinct.

> [The cruelty law] serves to define the outer boundaries of acceptable human conduct toward animals. By contrast, the unlawful hunting statute serves to enforce the authority of the Commission in defining the manner and conditions of lawful hunting and fishing in New Mexico and to ensure that hunting and fishing in New Mexico is carried out in [order] ... to provide an adequate and flexible system for the protection of the game and fish of New Mexico and for their use and development for public recreation and food supply, and to provide for their propagation, planting, protection, regulation and conservation to the extent necessary to provide and maintain an adequate supply of game and fish within the State of New Mexico.[167]

However, the court held that the legislature intended the fish and game laws to pre-empt the anti-cruelty law with respect to conduct contemplated under the fish and game laws. The fish and game laws implement a system under which the animals may be "use[d] and develop[ed] for public recreation and food supply," and delegate to the Commission the power to "authorize or prohibit the killing or taking of any game animals, game birds or game fish of any kind or sex" and the power to regulate "the manner, methods and devices which may be used in hunting, taking or killing game animals, game birds and game fish."[168] The game laws authorize the use of snares to trap certain fur-bearing animals which could be considered cruel. For example, as in this case, the manner of death for the two snared deer – strangulation and either starvation, dehydration, or fatigue – is not atypical for a snared game animal. Furthermore, approximately 25–30 percent of all deer shot are shot in the abdomen. Both results are likely in violation of the anti-cruelty laws but not addressed under the game laws. Thus, even if the legislature intended that wildlife generally be protected under the anti-cruelty laws, in the context of hunting, the fish and game laws would pre-empt application of the cruelty law.[169]

Compare this analysis to states where the courts have held that the anti-cruelty law covers wild animals. In such states, nuisance wildlife control operators have challenged their prosecution for cruelty when allegedly complying with the nuisance wildlife laws. For example, in *Connecticut v. Lipsett*,[170] the nuisance wildlife law required nuisance wildlife control operators to kill "nuisance" raccoons within 24 hours of their capture. The defendant control operator was charged with animal cruelty for drowning two trapped raccoons in their traps.[171]

First, the defendant argued that the cruelty statute was unconstitutionally vague, as it gave insufficient notice of what constituted "cruelty."[172] The court quoted *Webster's New World Dictionary*'s definition of cruel as "deliberately seeking to inflict pain and suffering; enjoying others' suffering; without mercy or pity" and "causing, or of a kind to cause, pain, distress, etc." and cruelty as the "willful infliction of physical pain or suffering upon a person or animal, or of mental distress upon a person."[173] The court held that "to violate the statutory prohibition of 'cruelly kills' as applied to wild animals requires an intentional or wanton infliction of pain or physical torture beyond that involved in the act of killing," and given the common understanding of these terms, is not overly vague.

The defendant, having lost that argument, then argued that drowning the raccoons did not violate the game statute and that therefore he could not be prosecuted for cruelty. The nuisance wildlife statute merely requires that he kill (not *humanely* kill) the trapped wildlife, and trappers can lawfully drown raccoons under the wildlife law. In fact, the state's trapper education course describes the proper use of drowning traps as the preferred method of trapping for raccoons. The prosecution countered that intentional drowning is by definition cruel "because it unnecessarily inflicts 10–12 minutes of distress, panic and struggling for air as the animal dies."[174] Moreover, several other means of euthanasia are much swifter and/or less painful.

The court held that the state lacked sufficient evidence to establish that the defendant was guilty of cruelty. The court stated that the nuisance wildlife statute does not require euthanasia but rather only the destruction of the animal. Since "fur bearing wildlife can lawfully be drowned by trappers for profit or sport, subject to certain regulations, without violating the statute forbidding cruelty to animals, ... drowning a nuisance raccoon by a nuisance wildlife control operator who is compelled by law to promptly destroy the trapped animal cannot be criminal or a 'cruel' killing in violation of the animal cruelty law."[175] The court held that the state failed to meet its burden beyond a reasonable doubt that drowning is per se cruel and that the defendant had any cruel intent. The court made clear that this decision was "strictly limited to the facts of this case which involved raccoons trapped by a nuisance wildlife control officer which were required to be destroyed because of the threat of rabies determined to exist in Connecticut in 1996."[176]

Compare the following case where an Indiana court upheld an animal cruelty charge in *Boushehry v. Indiana*,[177] when the defendant shot at two geese (during closed season), killing one and wounding the other. The defendant caught the wounded goose, grabbed it by the neck, found a knife, slit its throat, and threw the goose to the ground. The goose flopped around before becoming still. The defendant was charged with eleven criminal offenses, including animal cruelty, and found guilty of seven.[178] The Indiana anti-cruelty law prohibits "knowingly torturing or mutilating" an animal. The court found that the killing of the first goose was not cruel since the goose died instantly. However, the wounded goose lived after it was shot and slitting its throat is "mutilation in its plain, or ordinary and usual, sense."[179] It was clearly covered under the anti-cruelty statute. Although the defendant claimed he slit the goose's throat to put it out of its misery and that he did not intend to be cruel, the appellate court held that such a finding was for the trier of fact, especially since it was the defendant who caused the misery in the first place. Perhaps the fact that the defendant shot the geese during closed season and thus in violation of the game laws as well explains the court's willingness to prosecute him under the anti-cruelty law.

Hunting and trapping

In recent years, states have begun to outlaw certain types of hunting or trapping practices. In the 1990s, citizen initiatives to ban certain techniques had been successful to prevent the cruel killing of wildlife.[180] The trapping initiatives targeted leg-hold and other body-gripping traps. The animal advocates arguing for the initiatives claimed that these trapping methods are cruel because they kill indiscriminately any animal that is caught. Moreover, the pain is so unbearable while in a leg-hold trap, the animal will often chew its leg off to escape. Further, the animals who do not die suffer from severe injury while those who do die suffer from a prolonged and painful death. Four states, Arizona, Colorado, Massachusetts and California,[181] all passed bans on such traps. Other states have not been as successful. For example, Alaska had the opportunity to ban wolf snaring – the use of wire loops anchored to trees and designed to catch and strangle wolves (as well as other animals).[182]

Although snaring arguably is inhumane, opponents argued that banning wolf snaring endangers tradition and native way of life. At the polls, tradition won.

California also banned the use of poisons for predator control. Opponents argued that banning poisons for pest control endangers public health and safety as well as livestock and other wildlife since poison is the primary means of pest control for animals that carry disease or kill livestock or endangered species. However, the proponents won, arguing that the poisoned animals suffer an agonizing death and may poison other animals that feed off the poisoned dead animal.

Hunting initiatives tend to focus on hunting methods that could be characterized as unfair and cowardly, in addition to being cruel. Hunting methods include bear-baiting (placing materials out to attract a bear and then killing the bear from a blind spot) and bear-hounding (using dogs to chase a bear up a tree and then shooting the bear). These initiatives were successful in Colorado, Massachusetts, Oregon, and Washington, but defeated in Idaho and Michigan. Interestingly, bow hunting, although illegal in the United Kingdom, is legal in most states, and several states allow crossbow hunting as well.[183] This is so even though a report on bow hunting summarizing 24 studies across the United States concluded that bow hunting is inhumane and wasteful.[184] The report states:

> The possibility of a deer being impaled by a broadhead arrow and then dying instantaneously is extremely slight. Wounding and crippling losses are inevitable. Every one of these studies has concluded that for every deer legally killed by bowhunters, at least one or more is struck by a broadhead arrow, wounded, and not recovered. The studies indicate an average bowhunting wounding rate of 54%, with the shots per kill averaging 14.[185]

On the other side of the ledger, some states have attempted to amend their constitutions to add hunting, fishing and the taking of game as a constitutional right of its citizens. In Minnesota, the amendment that passed reads: "Hunting and fishing and the taking of game and fish are a valued part of our heritage that shall forever be preserved for the people and shall be managed by law and regulation for the public good."[186] Although the word "right" was removed from the original language, the effect of the provision is likely the same. Opponents of the amendment feared that such a provision would prevent the banning of leg-hold traps and other inhumane means of taking an animal and argued that the state's wildlife belongs to all citizens, not just the hunters and trappers.

Moreover, the amendment was unnecessary as the activities were not threatened but were supported by a large portion of the electorate. Finally, creating constitutional protections for recreational activities while not providing protection for basic life necessities such as food, water, shelter and medical care for humans, set bad constitutional law precedent. Minister Richard Gist published an editorial arguing against the amendment and stated eloquently that

> [c]ompassion struggles, always. Indifference brushes it aside, selfish pursuit forces it on the defensive. When cruelty is given free reign, compassion becomes muted, even derided ... I suspect that many in the world of trapping fear that my children's generation may be even more compassionate than their father's. The only way to remove the issue from future debate and legislative regulation is to write it in stone, now. But we're talking about our Constitution, which is not a vehicle for special interest groups to manipulate the future.[187]

A brief comparison to Australasia

General laws

As in the United States, the category of wild animals is broadly defined throughout Australasia as fauna, feral animals, wildlife, game animals, native animals, and pests.[188] Further, the legal treatment of wildlife in Australasia is quite similar to that in the United States. Steven White has noted that, in Australasia, the application of cruelty codes to wildlife is quite limited. For example, some animal welfare laws govern only "owners" of animals, which by definition excludes wild animals. He notes that this is not surprising as the laws also impose a duty of care to the animals under a person's control. However, the cruelty laws do govern "persons" not "owners," and thus arguably protect wild animals from intentional cruelty. Nevertheless, White explains that, as a practical matter, these laws often do not protect wild animals either because no one has standing to enforce the law or the courts fail to recognize that particular practices against wild animals are cruel.[189] For example, in a 2007 case seeking an injunction against the aerial shooting of goats and pigs in nature reserves in New South Wales on the basis that it would violate the cruelty codes, the judge held that there was insufficient evidence that the animals suffered unnecessarily after being shot.[190] Moreover, there are a number of legal defenses to prosecution under the cruelty codes including: (1) for the control of introduced wild animals (for example, pests) if the practice is "generally accepted as usual and reasonable"; (2) if the conduct is in compliance with codes of practice[191] governing the exploitation or farming of wild animals that generally would not otherwise satisfy the cruelty laws; and (3) if the action is done in accordance with conservation acts.[192]

The Kangaroo Codes

An example from Australia demonstrates the unique difficulties of protecting wild game animals from cruelty. When one thinks of Australia, one envisions the kangaroo – the majestic animal who leaps to cover wide terrain with the female carrying and protecting her young in her pouch. When the Commonwealth's Minister for Environment and Heritage approved the New South Wales Commercial Kangaroo Harvest Management Plan 2007–11, the Wildlife Protection Association of Australia brought a suit to challenge the approval, claiming that the Minister improperly assumed that compliance with the Code of Practice for the Humane Shooting of Kangaroos would satisfy the humane treatment requirement of the law. The Association argued that the Minister's assumption was not valid since instantaneous death of the mother does not always occur and orphan kangaroos are often left to die of starvation (young-at-foot[193]) or are killed by a severe blow to the head using a tow-bar (joeys[194]). The court upheld the Minister's assumption stating:

> Any management plan that involves commercial killing of free-ranging animals will involve a risk that perfection will not always be achieved. What is required is that the Plan achieve as near to perfection as human frailty will permit. We are satisfied that the system of accreditation, licensing, and compliance management achieves that object.[195]

Regarding the young kangaroos the court stated:

> Again, it may be accepted that there will be a very small number of instances where young at foot die [due to staving or being taken by predators], but we do not regard that fact, even in combination with the instances where an instantaneous killing of the adult is not possible, as leading to the conclusion that the Plan does not satisfy the object of promoting the humane treatment of wildlife.[196]

Australia updated its Kangaroo Codes[197] effective November 2008. The goal of the Code is "to ensure that all persons intending to shoot free-living kangaroos or wallabies ... undertake the shooting so that the animal is killed in a way that minimises pain and suffering."[198] Thus, the Code states that "when shooting a kangaroo or wallaby,[199] the primary objective must be to achieve instantaneous loss of consciousness and rapid death without regaining consciousness. For the purposes of the Code, this is regarded as a sudden and humane death."[200]

The "Code is implemented through education and relevant ... legislation as appropriate. [However,] the requirements of this Code do not override State or Territory animal welfare legislation. A lack of knowledge of relevant State or Territory animal welfare legislation is no defence against prosecution for animal welfare offences."[201] In order to achieve the objective of humane shooting the Code specifications describe the firearms and ammunition that may be used, and require that the animals may not be shot from a moving vehicle, that they must be stationary, standing and clearly visible, that the shooter should target the brain and must be certain each animal is dead before targeting another. If the injured animal is sill alive, "every reasonable effort must be made immediately to locate and kill it before any attempt is made to shoot another animal," and shooters should avoid females "where it is obvious that they have pouch young or dependent young at foot."[202] In the case that such a female is shot, any dependent young-at-foot must be shot and the female's pouch thoroughly examined and any pouch young euthanized. Methods to euthanize young include a single forceful blow to the base of the skull to destroy functional capacity of the brain, stunning immediately followed by decapitation, or a single shot to the brain or heart.[203]

These regulations are among the most detailed regarding the shooting of wild animals. Of course for those who believe that the hunting of these animals is unjustified no matter how the activity is regulated, they give little comfort. However, are these regulations effective at protecting kangaroos from inhumane killing? Under the prior Code, it was estimated that "each year over 4% or 100,000 kangaroos are not killed humanely" by commercial hunters and an even greater proportion is likely by the non-commercial hunter. [204] Moreover, the Code itself is not law unless codified by the state. Further, even if the state incorporates the Code, it is perhaps even more difficult to enforce hunting laws than other animal protection laws. This is so because the number of offenses are broad, including

> not killing animals humanely, using the wrong method, killing the wrong species, hunting in protected areas, exceeding bag limits, and not retrieving and killing wounded animals ... Violations are likely not

to be reported even if witnessed by others and the likelihood of others witnessing violations is slight given they may often occur at night, on private land, in rural and remote areas, and for safety reasons, ... from a location not visible to another person other than a fellow hunter.[205]

Furthermore the agencies empowered to enforce the laws are under-funded and/or not properly equipped to regulate activities throughout the year and in remote locations. In sum, so long as hunting remains legal, protecting animals in the wild from cruelty remains very difficult.

Enforcement: public and private

The criminal and other public laws described are enforced by public entities – the police, animal cruelty officers, the game commission, and prosecutors. Unfortunately, in virtually all jurisdictions, the laws are inadequately enforced for a number of reasons. The primary reasons are a lack of resources and competing priorities. Many law enforcement agencies are under-funded and overworked. In many cases the same officers or lawyers enforcing the animal cruelty laws are also enforcing other criminal laws that involve human victims. The human cases always take priority. Moreover, the victims of animal cruelty have no voice, and rely on humans to uncover the cruelty and alert the appropriate officials. Thus, if the animal's owner is the abuser, it will be especially difficult to identify the cruelty much less investigate and prosecute the abuser. Animal advocates have attempted to capitalize on the demonstrated link between animal cruelty and human violence described above to justify allocating more resources to enforcing animal cruelty by noting the direct benefit to saving humans from violence as well.

A solution to offset the resource issue is to allow private entities to enforce civilly the criminal law.[206] In this manner, public monies are not used to subsidize enforcement efforts and the animals can escape the cruelty even if the perpetrator is not sanctioned. The State of North Carolina, for example, provides

a civil remedy for the protection and humane treatment of animals in addition to any criminal remedies that are available ... A real party in interest as plaintiff shall be held to include any 'person' as hereinbefore defined even though such person does not have a possessory or ownership right in an animal; a real party in interest as defendant shall include any person who owns or has possession of an animal.[207]

The remedy is a preliminary and/or permanent injunction granting the plaintiff the right to provide suitable care for the animal and, if necessary to protect the animal, to remove the animal from the defendant and take possession of the animal. If the plaintiff ultimately prevails against the defendant, the court may award the plaintiff costs of food, water, sheltering, and care of the animal while under plaintiff's care. Moreover, if warranted, the defendant's ownership in the animal may be terminated permanently and the defendant may be enjoined from future ownership of an animal for a specified period of time.

The North Carolina provision is quite broad with respect to potential plaintiffs. Other jurisdictions limit the plaintiff to a recognized humane organization. The fear is that broad standing provisions invite frivolous suits against defendants. However, the court system has independent mechanisms for deterring and punishing the filing of frivolous suits, minimizing this fear. Moreover, filing a suit is not costless. Plaintiffs pay with their money, time, and emotion. Since only an injunction is available, there is no monetary incentive to bring such suits. Thus it is highly unlikely that individuals would file any suit unless closely associated with the animal victim, much less file frivolous suits. Another concern is the potential to disrupt a criminal prosecution. That concern can be addressed by requiring that the plaintiff notify the prosecutor before filing a civil suit. The prosecutor then decides either to file criminal charges or to allow the plaintiff to proceed civilly. Such provisions allowing for private enforcement could significantly help abused animals. Even if the defendant is not criminally prosecuted, at least the animal will be protected from further cruelty.

3
Animal Welfare Laws

Introduction: animal welfare acts

Animal welfare laws regulate humans' use of animals. Humans use animals in a number of different ways, including for research, entertainment or exhibition, hobby, and food. Animal welfare is defined as an animal's state of well-being, or "maintenance of animals under conditions of space, environment, nutrition, and so forth, consistent with the physiological and social needs of the species" consistent with the Five Freedoms: freedom from (1) thirst, hunger and malnutrition, (2) discomfort, (3) pain, injury, and disease, (4) fear and distress, and (5) freedom to express normal behavior.[1] Arguably these laws should set standards beyond the mere prohibition of cruel practices already outlawed by the anti-cruelty laws. However, this may not necessarily be the case, depending upon the requirements of the particular human use being regulated. As we noted, the anti-cruelty laws often exempt certain communities of animals, such as animals used for food or research and/or practices; for example, "accepted agricultural practices." For these animals, the animal welfare laws provide the only legal protection available.

In the United States, the federal laws governing animal welfare are enacted by agencies, such as the United States Department of Agriculture (USDA) or the Fish and Wildlife Service (FWS), pursuant to legislative grants of power. Public enforcement is done by the same agencies with some limited opportunity for private enforcement as well. Generally, the sanction for violation of a regulation is loss of a license or permit, and/or the imposition of fines.

The primary federal statute that governs the treatment of animals is the Animal Welfare Act (AWA), first enacted in 1966 and amended several times since. The AWA regulates the care and handling of animals during transport, used in research or exhibition, and bred for sale as companion animals. The AWA was enacted pursuant to Congress' Commerce Clause powers, and consequently only regulates conduct that has a substantial effect on interstate commerce. Although this limits the application of the AWA significantly, it is nevertheless important to have a federal law to establish minimum national standards for the humane care and treatment of animals involved in interstate commerce. Upon enactment, Congress stated that the AWA is:

(1)   to insure that animals intended for use in research facilities or for exhibition purposes or for use as pets are provided humane care and treatment;
(2)   to assure the humane treatment of animals during transportation in commerce; and
(3)   to protect the owners of animals from the theft of their animals by preventing the sale or use of animals which have been stolen.[2]

The AWA governs dealers of animals. "Dealer" is defined as:

> any person who, in commerce, for compensation or profit, delivers for transportation, or transports, except as a carrier, buys, or sells, or negotiates the purchase or sale of, (1) any dog or other animal whether alive or dead for research, teaching, exhibition, or use as a pet, or (2) any dog for hunting, security, or breeding purposes, except that this term does not include –
> (i) a retail pet store except such store which sells any animals to a research facility, an exhibitor, or a dealer; or
> (ii) any person who does not sell, or negotiate the purchase or sale of any wild animal, dog, or cat, and who derives no more than $500 gross income from the sale of other animals during any calendar year.[3]

"Animal" is defined as:

> any live or dead dog, cat, monkey (nonhuman primate mammal), guinea pig, hamster, rabbit, or such other warm-blooded animal, as the Secretary may determine is being used, or is intended for use, for research, testing, experimentation, or exhibition purposes, or as a pet; but such term excludes (1) birds, rats of the genus Rattus, and mice of the genus Mus, bred for use in research, (2) horses not used for research purposes, and (3) other farm animals, such as, but not limited to livestock or poultry, used or intended for use as food or fiber, or livestock or poultry used or intended for use for improving animal nutrition, breeding, management, or production efficiency,

or for improving the quality of food or fiber. With respect to a dog, the term means all dogs including those used for hunting, security, or breeding purposes.[4]

These definitions exclude many persons who actually deal in animals as well as many animals. We will address the scope of these definitions and exclusions as we discuss the regulation of each category of animal use governed by the AWA.

Animals used for research

The AWA House Conference Report, discussing animals used in research, stated: "We have diligently tried to bring back to the House an effective bill which will codify the noblest and most compassionate concern that the human heart holds for those small animals whose very existence is dedicated to the advancement of medical skill and knowledge while at the same time still preserving for the medical research professions an unfettered opportunity to carry forward their vital work on behalf of all mankind."[5] It is important to note that while the motivation is compassion for the animals, the goal is to preserve *unfettered* research efforts to benefit *man*kind.

Goals and functions

The AWA requires that all dealers who supply animals to research facilities be licensed and that research facilities only purchase dogs or cats from a licensed dealer.[6] This provision is designed to protect owners from the theft of their companion animals for sale to research facilities. A primary function of the AWA is to provide for the humane care and treatment of animals used in research. The AWA requires that the Secretary of the USDA "promulgate standards to govern the humane handling, care, treatment, and transportation of animals by dealers, [and] research facilities," including minimum requirements for food, water, shelter, and veterinary care for all animals, and "for exercise of dogs, as determined by an attending veterinarian in accordance with general standards promulgated by the Secretary, and for a physical environment adequate to promote the psychological well-being of primates."[7]

The Code of Federal Regulations (CFR) contains detailed specifications for the handling, care, treatment and transportation of animals.[8] The specifications are divided among a number of different groups of animals. For example, there are specifications for "Dogs and Cats," for "Guinea Pigs and Hamsters," for "Rabbits," for " Nonhuman Primates," for " Marine Mammals," and for all other "Warmblooded Animals." For each group the regulations define parameters governing the facilities in which the animals are kept, proper feeding, watering, transportation, and compatible grouping for the animals.[9]

One significant problem with the AWA is that it protects less than 10 percent of all animals used in research.[10] Originally the AWA defined "animal" as "dogs, cats, monkeys, guinea pigs, hamsters, and rabbits." In 1970, the definition was amended to add any "warm-blooded animal, as the Secretary may determine is being used, or is intended for use, for research, testing, experimentation, or exhibition purposes, or as a pet," but excluded "horses not used for research purposes and other farm animals ... used or intended for use as food or fiber ..."[11] The Secretary subsequently determined that the 1970 definition *excluded* birds, mice and rats – representing over 90 percent of all animals used in research – and issued a rule to that effect.

Animal advocacy groups filed suit to challenge the rule. Challenging agency decisions is difficult as the courts give great deference to agency interpretation of Congressional statutes and to the rules they enact. Challengers must prove to the court that the agency's rule or decision is "arbitrary, capricious, an abuse of discretion, or otherwise not in accord with the law pursuant to 5 USC 706(2)(A)"[12] in order to prevail.

In 1992, in a surprising decision, Judge Richey of the United States District Court for the District of Columbia held the agency rule that excluded birds, mice and rats from any "warm-blooded animal, as the Secretary may determine is being used, or is intended for use, for research, testing, experimentation, or exhibition purposes, or as a pet"[13] was arbitrary and capricious and without basis in law. In a thorough analysis of the statutory language, legislative history, and underlying statutory policy, he explained his decision as follows. First, as a matter of construction, Judge Richey stated that the plain language of the statute clearly includes birds, mice, and rats because the discretion granted the Secretary under the AWA relates to the determination of whether the animal "is being used or is intended for use, for research," not to the definition of "animal" itself.

Second, as a matter of legislative intent, the agency, arguing in support of its rule, referenced a "single sentence from the Committee report which state[d]: 'Under this section of the bill, it would be expected that the Secretary would designate additional species of those animals not previously covered *as permitted by available funds and manpower.*'"[14] The agency argued that inadequate resources – for example, funds and manpower – were available to cover such animals because of the number of animals involved and the approximate cost of regulation. Judge Richey, however, found that the sentence in the report "refer[s] to the Secretary's power to designate additional species for the expenditure of Department funds and manpower, not for coverage by the Animal Welfare Act, which is mandated by the language of the Act."[15] Moreover, he stated that inclusion of these animals need not require the expenditure of significant resources as many of the requirements are self-implementing by the regulated facility.

Third, as a matter of policy, the agency's interpretation does not further the goals of the AWA to provide for the humane care and treatment of animals used in research. Judge Richey explained that the agency only considered the costs of promulgating and enforcing regulations for the treatment of birds, rats, and mice, and not the associated benefit to the animals of the regulations. Such one-sided analysis is antithetical to the express purpose of the AWA to protect animals used in research from inhumane treatment. Judge Richey explained that including these animals

> would send an important message to those responsible for their care – that the care of these animals is something for which they are legally accountable and is an important societal obligation ... [instead of the message now being sent] that the researchers may subject birds, rats and mice to cruel and inhumane conditions, that such conduct is sanctioned by the Government and has no legal consequences.[16]

The judge further chastised the agency in a footnote, stating:

> The agency's argument that it lacks the resources to implement these regulations might be more convincing if the agency sought more resources to pursue its mandate. In fact, the plaintiffs have shown that the agency intentionally sought funding *decreases* and one year requested that its Animal Welfare Program be *eliminated* ... This evidence suggests that the agency may have lost sight of its Congressional mandate under the Act. A member of the President's Cabinet charged with executing the law should not be a prisoner of his own bureaucracy and allowed to argue that his own failure to request funding to comply with an Act of Congress is a proper excuse for his failure to pursue his statutory obligations.[17]

This decision was a victory for the plaintiffs and, more importantly, for the birds, rats, and mice used by research institutions. However, the victory was short-lived. On appeal, the D.C. Circuit held that the plaintiffs lacked standing to challenge the agency's rule-making and dismissed the case.[18] Four years later, the D.C. Circuit revisited the standing doctrine[19] and a second suit was filed to raise the identical challenge. After the court denied the agency's motion to dismiss for lack of standing, the parties settled the case, and the agency agreed to revise the regulations to include these animals. However, soon thereafter, Congress intervened and amended the AWA definition of "animal" to exclude birds, mice and rats.[20] Senator Jesse Helms introduced the amendment, claiming it was necessary so that "none of the important work taking place in the medical research community will be delayed, made more expensive, or be otherwise compromised by regulatory shenanigans."[21] He argued that by including these animals, the "USDA will force researchers out of the laboratory to spend their time filling out countless forms for yet another federal regulator."[22] The National Association of Biomedical Research lobbied for the amendment, stating that "new animal care standards are unnecessary, costly and would delay cures of diseases."[23] Although there was significant opposition to the amendment, including not only animal welfare groups, but Institutional Animal Care and Use Committee (IACUC) members charged with enforcement of the AWA, and Colgate-Palmolive and Procter & Gamble, two companies that test their products on animals; the amendment passed.[24] This amendment was a devastating blow to the animals and to society's obligation to provide for the humane care and treatment of animals used in research.

To provide for the humane treatment of the other animals used in research, the AWA provides that the Secretary promulgate standards for their humane treatment, and specifically to provide minimum requirements "for exercise of dogs, as determined by an attending veterinarian in accordance with general standards promulgated by the Secretary, and for a physical environment adequate to promote the psychological well-being of primates." Pursuant to these specific requirements, the Secretary promulgated the following standards: for dog exercise, "The frequency, method and duration of the opportunity for dog exercise shall be determined by the attending veterinarian and, at research facilities, in consultation with and approval by the Committee;"[25] and for "the psychological enrichment of non-human primates, many of whom have the distinctive qualities of human beings ... [T]he regulated entity [shall] devise a plan to foster [the psychological enrichment of non-human primates] 'in accordance with the currently accepted professional standards as cited in appropriate professional journals or reference guides, and as directed by the attending veterinarian.'"[26]

The regulations require that each plan address certain topics, including social grouping and environmental enrichment but fail to include minimum requirements or standards and instead rely on generally accepted professional standards.

A suit was filed against the agency challenging these regulations in the D.C. District Court. Judge Richey again heard this case and found for the plaintiffs. He held that the regulations were "arbitrary and capricious."[27] Judge Richey explained that by providing no minimum requirements the regulations leave total discretion to the on-site veterinarian at the research facilities. This, in turn, leaves the AWA open to a wide variety of interpretations, with compliance dependent upon the good faith of the regulated entity. He explained that this was contrary to the AWA's express language that the agency promulgate standards of minimum requirements for dogs and primates. The agency attempted to persuade Judge Richey that Congress' use of the word "standards" was synonymous with "regulations" and specifically allowed reference to veterinarian discretion. Judge Richey was not convinced. He explained that the command for minimum requirements does not simply refer to the promulgation of regulations. He stated: "to believe the Government in this regard would be to assert a belief in the tooth fairy."[28]

Judge Richey found even more persuasive the fact that

> in their initial proposed [dog exercise] regulations, [the agency] concluded that '[t]he consensus of [its] veterinarians with training and experience in the care of dogs is that 30 minutes of daily exercise is a reasonable minimum for maintenance of a dog's health and well being.' The Defendants later argued that minimum provisions would not take into account any variation among the type of dogs involved and would be too restrictive of diverse facilities.[29]

Similarly, the regulations governing primates were inconsistent with the agency's own original judgment on the psychological enrichment of non-human primates, including social grouping and cage sizes. The agency had determined that

> the social deprivation is psychologically debilitating to non-human primates and that group housing for non-human primates was the best way to avoid this problem, [and] ... that it is their expectation that non-human primates will be housed in groups. However, despite these findings, nothing in the regulations requires group housing nor

is any explanation given as to why group housing is not required under the regulations.[30]

Apparently, the costs to the regulated facility would be too great. Judge Richey stated: "These considerations may well be based more on the almighty dollar than the welfare of animals, and do not excuse ... nor authorize [the agency] to ignore the Act's plain mandate to set minimum requirements for the humane care and treatment of animals."[31] He then held that the regulations were arbitrary and capricious. Unfortunately, as before, the victory for the dogs and primates was short-lived. On appeal, the D.C. Circuit found the plaintiffs lacked standing.[32]

Rarely, if ever, has a federal judge given the care and consideration to research animals' interests and the obligation of the Secretary to implement the terms of the Animal Welfare Act as Judge Richey did in these two cases. Judge Richey was not afraid to speak his mind and interpret the statute and Congressional intent in a light favorable to the animals whose interests were supposed to be served. For example, in this case he stated: "Article III courts were not created by our founding fathers to rubber stamp such failures to act over indefinite periods while bloated bureaucrats contend with each other and the special interest groups who transfer their efforts from the Legislative Branch to the Executive Branch, after a bill has passed."[33] He also said: "'A dog is man's best friend' is an old adage which the [agency] has either forgotten or decided to ignore. Hopefully the new Secretary will ensure that the bureaucracy he inherits and the special interest groups with which he must contend will be forced to remember this sentiment and comply with the law."[34] Judge Richey died in 1997, but will be remembered by animal advocates for his stand against the USDA and their implementation of the AWA.

Procedures for implementing

Each research facility that intends to use live animals for research or testing must register with the Secretary and agree to comply with the regulations and standards of the AWA. The regulations require that each research facility have an Institutional Animal Care and Use Committee, comprised of a Chairman and at least two additional members, including a Director of Veterinary Medicine, and at least one member not affiliated with the research facility, to assess the facility's animal program, facilities and procedures. The IACUC is responsible for ensuring that all activities meet a number of specific criteria, including that: (1) procedures will avoid or minimize discomfort, distress, and pain to the animals; (2) the principal investigator has considered alternatives to procedures that may cause more than momentary or slight pain or distress to the animals, and has provided a written description of the methods and sources used to determine that alternatives were not available; (3) "procedures that may cause more than momentary or slight pain or distress to the animals will be performed with appropriate sedatives ... unless withholding such agents is justified for scientific reasons;" (4) animals that would otherwise experience severe or chronic pain or distress that cannot be relieved will be painlessly euthanized at the end of the procedure or, if appropriate, during the procedure; and (5) animals' living conditions will be appropriate for their species.[35]

The three Rs

The regulations arguably are designed to implement the 3Rs – Reduce, Refine and Replace. The 3Rs is the primary model for tackling the issue of animals and their use in research.[36] "Reduce" refers to minimizing the number of animals used in a research project through improved design of the experiments; "Refine" refers to minimizing animal suffering in a given project through the use of anesthesia or other means; "Replace" refers to the adoption of non-animal alternatives for the project. Arguably, the goal is to phase out the use of live animals in research in most cases.

When Congress amended the AWA in 1985, it stated:

Congress finds that ... methods of testing that do not use animals are being and continue to be developed which are faster, less expensive, and more accurate than traditional animal experiments for some purposes and further opportunities exist for the development of these methods of testing, [and] measures which eliminate or minimize the unnecessary duplication of experiments on animals can result in more productive use of Federal funds.[37]

How does the AWA implement the 3Rs? To "refine" the statute states that animal pain should be minimized and requires that the Secretary establish regulations that govern the use of analgesics when a practice might cause the animal pain unless it is scientifically necessary to withhold pain medication. To "reduce and replace" the AWA requires that the main investigator consider alternatives to the use of live animals and that the Secretary establish an information service to provide information on methods that "reduce or replace animal use; and minimize pain and distress to animals."[38] Moreover, the IACUC is to review the research performed at their institutions for minimizing pain and considering alternatives.

Has the AWA been successful in satisfying the 3Rs? Unfortunately, not. First, and most importantly, the AWA does not address the most fundamental question – whether the purpose of the specific research project justifies *any* animal suffering.[39] The AWA states that the Secretary "cannot promulgate rules ... with regard to the performance of actual research or experimentation by a research facility as determined by such research facility,"[40] and its own IACUC cannot challenge the "design, performance, or conduct of actual research or experimentation."[41] Second, as noted above, over 90 percent of all animals used in research are exempted from coverage under the AWA. Third, "few attempts have been made to assess the quality of design of animal experiments"[42] and the researchers' claims of "scientific necessity" are virtually never challenged. Further, the only requirement is that the researcher considers alternatives, but since the IACUC cannot interfere with the performance or design of the project, there is no means for requiring that even proven alternatives be used. Finally, the ultimate catch-22 is that the effectiveness of alternatives is based on a comparison to live animal experimentation and thus, in the context of new technology, the 3Rs is virtually meaningless until after live animal protocols have been developed and tested.[43]

Enforcement

The primary enforcement of the AWA is public enforcement by the USDA since the AWA does not provide for private rights of action. The only means for advocates to challenge the Secretary's regulations is through the Administrative Procedures Act (APA) and standing to bring these challenges is very limited. Moreover, even if a proper plaintiff is found, the courts give extreme deference to the agency.

To the extent that there are meaningful regulations promulgated by the Secretary, are these regulations followed by the research facilities? Animal Care (AC), a subdivision of the USDA within the Animal and Plant Health Inspection Service (APHIS), is charged with enforcing the AWA. Research facilities are to be inspected annually. The AWA provides only for civil penalties for violations by researchers of no more than $10,000 for each violation. In determining the penalty, AC evaluates the size of the facility, the gravity of the offense, the good faith of the facility, and the facility's history of AWA compliance. In a 2005 report, the USDA's Office of the Inspector General found enforcement of the AWA inadequate.[44] The executive summary of the audit report highlighted the lack of enforcement, the minimal fines, failure to follow the 3R protocols, and ineffective tracking of violations, prioritizing of inspections, and collection of penalties.[45] This assessment from the agency's own inspector presents a rather grim picture of the enforcement of the AWA.

In 2009, the agency settled a case that may improve the situation slightly.[46] Transparency is a useful tool to promote greater compliance with the law and to protect animals from inhumane treatment. The reason is twofold: the public is informed of the actual treatment of the animals, which in turn generates public interest to protect the animals, and the actors do not want to be discovered breaking the law and/or treating the animals inhumanely. The Humane Society of the United States (HSUS), in 2005, filed suit against the USDA claiming violation of the Freedom of Information Act (FOIA) for failing to provide numerous documents they had requested. Under the AWA, research facilities must submit annual reports and include the number and species of animals used, the pain and distress category of research conducted, and explanations for failure to provide pain medication to the animals. Filing FOIA requests is a tedious process and it generally takes a long time to receive the information from the government. In this particular case, HSUS had been requesting information on research facilities and their use and treatment of animals since 2001.[47]

In May 2005, the USDA agreed to resume posting the annual reports on their website after having pulled all reports from the website upon the request of the animal research labs.[48] However, several issues remained, including missing reports and large amounts of redacted information in the reports. Thus, litigation continued. During the course of litigation, the "USDA ... disclosed it never received mandatory annual reports about research animals' 'pain and distress' from 81 animal research facilities over the course of five years."[49] Finally, in 2009, the USDA settled with the HSUS. The settlement requires that the USDA post all animal research facility annual reports, including pain and distress data, online in a timely manner. Moreover, the USDA agreed to reduce the amount of information redacted before releasing the reports to the public.[50] The hope is that with greater public access to such information, the number of animals used in research will be reduced and their welfare improved.

However, until adequate resources are allocated to enforcement, the lack of enforcement documented by the inspector general likely will not improve. Quite interestingly, the USDA often requests *less* money than Congress budgets for enforcement of the AWA. This suggests that the USDA has little, if any, interest in enforcing the AWA. In 2007, some 110 inspectors handled the over 10,000 licensed facilities governed under the AWA. The amount appropriated in the budget for fiscal year 2007 was

$20 million, a very small amount compared to the total budget of the USDA of $92.8 billion.[51]

Do the IACUCs effectively govern research projects? As the 2005 audit report found, some IACUCs are not effectively monitoring animal care activities or reviewing protocols. The IACUC governs from within the research institution and only one member must be selected from "outside" the facility – in other words, not affiliated with the research facility. The political reality is that there is significant pressure on the IACUCs not to question the research protocols. Although the IACUCs represent the public interest, the reality is that they can be easily "captured" by the institutional interests. One of the most striking ironies concerning enforcement of the AWA is that courts have held that private individuals lack standing under the APA to enforce the AWA precisely because the IACUCs exist to protect the public interest, even when the person filing suit was himself a member of an IACUC.

A case study: standing and the AWA

In the United States, to be recognized by the court, a person or entity must have standing. To have standing, two requirements must be satisfied – an Article III constitutional requirement and a prudential statutory requirement. Article III grants courts jurisdiction over a "case or controversy." The U.S. Supreme Court has interpreted "case or controversy" to mean that the plaintiff suffered (1) an injury-in-fact (2) that is fairly traceable to the defendant's actions, and (3) redressable by the relief requested.[52] Prudential standing requires that the plaintiff fall within the "zone of interests" of the statute allegedly violated by the defendant. When filing a claim to enforce the AWA, the plaintiff must seek standing under the APA because the AWA does not grant a private right of action itself. The APA grants standing to any person "adversely affected or aggrieved by agency action within the meaning of a relevant statute."[53] The relevant statute in this case is the AWA.

The rationale behind the courts' constitutional limitation to decide "cases or controversies" is to maintain a separation of powers between the courts and Congress. The courts are granted the power to adjudicate – to decide disputes among parties and redress the alleged wrong. Congress is granted the power to legislate – to establish public policy through the creation of law. Thus, the elements of the standing doctrine are designed to define a "dispute" that falls within the courts' authority and expertise and to prevent courts from overreaching and infringing on the legislative power of Congress to establish policy. To define the proper party-plaintiff, the plaintiff must present an injury that is actual, direct, concrete and particularized, not abstract, uncognizable, conjectural or hypothetical. The proper defendant must be legally liable to the plaintiff and have caused the plaintiff's injury, and the remedy requested must be available through the court. Because the definition of "case or controversy" is rooted in a traditional characterization of the court's role – handling specific disputes between specific entities – with little or no overarching effect on others or public policy generally, it is often difficult for animal protection lawyers to find a proper party-plaintiff. Animal protection lawyers, as most all public interest lawyers, seek to use the courts to change the status quo and achieve substantive reform. Such use of the courts approximates, to some degree, legislative activity; the parties are seeking to use the counter-majoritarian courts instead of the majoritarian legislature to seek change of policy through legal reform for minority interests.

The Supreme Court has not addressed whether animals have constitutional standing to file a suit on their own behalf under the AWA or any other statute that grants them protection, such as the Endangered Species Act. However, the Ninth Circuit in *The Cetacean Community* v. *Bush*[54] held that nothing in the constitution limits the filing of claims in federal court to humans. The court noted that animals do have legal rights under a variety of statutes – both federal and state. The court explained that although animals themselves cannot file suit, there is no reason why they cannot be represented by an appropriate agent, as is done in cases filed on behalf of corporations, juridically incompetent humans, partnerships, or ships.[55] Thus, as long as the requirements of a "case or controversy" are met, the fact that the plaintiff is an animal would be irrelevant.

Prudential standing is a matter of statutory interpretation and legislative intent. Even if the plaintiff has constitutional standing to seek relief, the plaintiff also must demonstrate that Congress intended for him or her to obtain relief under the relevant statute. In other words, the plaintiff must fall within the "zone of interests" of the statute. The court has held that one who suffers a loss to an economic, professional, competitive, aesthetic, recreational, or informational interest caused by the violation of a statute may fall within the zone of interest of said statute.[56] Since the AWA was enacted to provide minimum protection to animals, the animals themselves clearly fall within the zone of interest of the AWA but no private right of action exists. Instead, as explained above, plaintiffs must use the APA to seek standing to raise AWA claims. However, animals lack prudential standing under the APA, as well as animal protection statutes that grant a private right of action, such as the Endangered Species Act (ESA). The reason is that the statutes expressly grant a private right of action to a "person" defined as "an individual, partnership, corporation, association, or public or private organization other than an agency."[57] Although, arguably, animals could be included within the term "individual," the courts have not so held. Thus, an animal is not a "person" under the statutes and consequently does not have standing to bring claims under statutes that purport to protect them.

Accordingly, in order to file suit, the animal protection lawyer must find a "person," an individual human and/or organization, who falls within the "zone of interest" of the AWA. This is difficult. Often several organizations and individuals file suit jointly, hoping at least one will survive a standing challenge inevitably raised by the defendant. Let's see how and why the cases decided by Judge Richey, discussed above, were dismissed on appeal for lack of standing.

In the case challenging the exemption of birds, mice, and rats from the definition of "animal" under the AWA, four plaintiffs filed suit: Dr. Patricia Knowles, William Strauss, the HSUS, and the Animal Legal Defense Fund (ALDF).[58] Knowles was a psychobiologist who had worked in research labs using live mice and rats. She alleged that the agency's failure to include these animals under the definition "rendered her 'unable to effectively control the care and treatment these institutions afforded the rats and mice she used'; that 'the inhumane treatment of these animals will directly impair her ability to perform her professional duties as a psychobiologist'; and that 'she will be required to spend time and effort in an attempt to convince the facility of the need for humane treatment.'"[59] The court held that since Knowles was not currently using animals in her research, her alleged injuries were not imminent. In reply, Knowles argued that to further her professional goals she would have to perform research on live mice and rats in the future, and thus will be subjected to these injuries at that time given her prior experiences. However, since she did not state a specific time frame for the work and she was not required to use animals but instead would choose to do so, the court found that her injuries were conjectural. Thus, she lacked constitutional standing.[60] Judge Williams dissented and would have held that her future plans to experiment on mice and rats, given her professional interests, were sufficiently impending to qualify as imminent. Moreover, he explained her injuries met the other requirements for constitutional standing as well. The delay and possible frustration of her professional research plans were clearly concrete economic and professional losses. Moreover, she would suffer aesthetic injury if forced to see the inhumane treatment of the mice and rats while working in a lab with no protocols to protect the animals. These injuries would be remedied if she prevailed because the rats and mice would be covered under the AWA and protocols would be in place to protect them from inhumane treatment, allowing her to pursue her research plans free from harm.[61]

The second plaintiff, Strauss, was an attorney who was an IACUC member representing the public's interest in the proper treatment of animals in the lab. He alleged that "the agency's failure to promulgate standards governing the humane treatment of birds, rats and mice has left him 'without relevant guidance' in evaluating the treatment and conditions of those animals." As a result, he cannot adequately perform his statutory duties as a member of the committee.[62] Moreover, the lack of regulations for birds, rats, and mice hampered "his ability to ensure that those animals' living conditions are appropriate for their species, as the Act allegedly requires."[63] The court found that such injuries were not cognizable, stating:

> His suit amounts to nothing more than an attempt to compel executive enforcement of the law, detached from any factual claim of injury. That type of suit has no place in the federal courts, for '[v]indicating the public interest (including the public interest in government observance of the Constitution and laws) is the function of Congress and the Chief Executive.'[64]

Note that this finding is rather odd given that the AWA tasks the IACUC, of which he is a member, with this function.

The two organizations, the HSUS and ALDF, argued informational standing. They argued that the exclusion of birds, rats, and mice from the definition of "animal" hampered their ability to gather and disseminate information to their members and the public on their treatment by the research labs. The court held that while their injury met Article III requirements, it did not fall with the "zone of interests" of the AWA. The court held that "an organization must show more than a general corporate purpose to promote the interests to which the statute is addressed. Rather it must show a congressional intent to benefit the organization or some indication that the organization is 'a peculiarly suitable challenger of administrative neglect.'"[65] Since the AWA specifically "entrust[s] to the [IACUCs] the functions of oversight and the dissemination of information, [this] precludes any inference that other private advocacy organizations are 'peculiarly suitable challenger[s] of administrative neglect.'"[66] The irony here is that the court deprives the organizations of standing because the IACUC is supposed to oversee the administration. However, a member of the IACUC, Strauss, who is entrusted to oversee the administration, was deprived standing as well.

In the case challenging the dog and primate standards, Dr. Fouts and Primate Pole Housing, Inc. (PPH) additionally sought standing.[67] PPH manufactures innovative primate housing systems and alleged that they are unable to sell their systems because the regulations fail to require group housing for the primates, as they should. Thus, there is no incentive for the research facilities to purchase their product. The court found, unsurprisingly, that PPH does not fall within the "zone of interests" of the AWA since the sale of primate housing products is not a goal of the AWA.[68]

Dr. Fouts asserted "that the regulations' 'vagueness' prevents him from establishing a plan for his research institute, and in particular for a chimpanzee housing facility now under construction, that he can be certain will pass USDA muster."[69] The court held this injury lacking for two reasons. First, Dr. Fouts himself will suffer no injury since his institution, not he personally, is required to comply with USDA regulations. Second, as Knowles above, the court held that even if he could demonstrate personal harm, it is too conjectural because he does not know whether their facility will comply with the standards. If they did, there would be no injury.

Chief Judge Mikva made an interesting observation in concurrence. He stated "[h]ad the public interest organizations and individuals challenging the Secretary's regulations alleged an interest in protecting the well-being of specific laboratory animals (an interest predating this litigation), I think appellees would have had standing to challenge those regulations for providing insufficient protection to the animals."[70] He suggested that such an interest would satisfy both constitutional standing and would place them in the zone of interest under the AWA, given the AWA's goal is to protect laboratory animals.

In 1996, ALDF along with individual plaintiffs, including Marc Jurnove, filed suit to challenge the agency's regulations regarding minimum standards to promote the well-being of primates being exhibited.[71] Judge Richey again reached the merits, set aside the regulations, and remanded the case to the USDA to promulgate new regulations in accordance with the AWA. The D.C. Court of Appeals, in banc, upheld Mr. Jurnove's standing to bring the case.[72]

Jurnove was trained in wildlife rehabilitation and worked for humane organizations. He testified that he had a familiarity with and love for exotic animals and often enjoyed seeing them in zoos and other parks near his home. During a one-year period he visited the Long Island

Game Farm Park and Zoo ("Game Farm") at least nine times and saw many animals, including primates, living under inhumane conditions. He explained that as a result he suffers extreme aesthetic harm and emotional and physical distress caused by the failure of the USDA to enact regulations to promote the psychological well-being of the primates he witnessed in isolation and without an adequate environment.[73]

The court held that he satisfied the requirements of constitutional and prudential standing. First, he demonstrated a "direct, concrete, and particularized injury to this aesthetic interest" by repeatedly visiting the zoo and observing the specific animals living under inhumane conditions.[74] The court stated that the fact that Jurnove may share this injury with others does not make his injury less particularized. Further, injury to an aesthetic interest in observing animals living under humane conditions is cognizable. Second, the existing regulations cause his injury because they allow such inhumane conditions to exist. If the USDA promulgated lawful regulations to promote the primates' well-being as required under the AWA, Jurnove's injury would be redressed in his future planned visits to the Game Park as such inhumane conditions would be prohibited.[75]

The five judges in dissent believed that the majority departed from then-existing constitutional standing aesthetic injury jurisprudence. They stated that although courts have held that "the desire to use or observe an animal species, even for purely aesthetic purposes, is undeniably a cognizable interest for purpose of standing," those cases "are limited to cases in which governmental action threatened to reduce the number of animals available for observation and study."[76] Jurnove's injury to his interest in viewing animals kept under humane conditions is purely subjective based solely on his own value preferences and lies outside the constitutional boundaries of a case or controversy. Thus, it is not cognizable.[77] Even if cognizable, the dissent argued that Jurnove did not establish that the agency regulations authorized the conditions he witnessed and therefore failed to establish causation. There are an infinite variety of things that the agency does not expressly regulate but that does not, in turn, mean that the agency authorizes them.[78] Moreover, even if the agency were to promulgate new regulations, there is no proof that Jurnove's aesthetic injury would be redressed given the purely subjective nature of his injury.[79] Thus, the dissent would have held that Jurnove failed to meet all three requirements for constitutional standing.

The majority also found that Jurnove met prudential standing and fell within the zone of interests of the AWA. First, the majority explained that prudential standing under the APA asks "whether the interest sought to be protected by the complainant is *arguably* within the zone of interests to be protected by the statute" and that "[t]his analysis focuses, not on those who Congress intended to benefit, but on those who in practice can be expected to police the interests that the statute protects."[80] The court stated that the purpose of animal exhibits is to educate and entertain the public and the AWA was enacted to ensure the public that adequate safeguards were in place to protect the welfare of animals. Moreover, although the AWA creates oversight committees for research facilities, there is no counterpart for animal exhibitions. Instead, Congress anticipated that concerned citizens attending the exhibitions would monitor compliance with the purposes of the AWA.[81] The dissents did not address the prudential standing requirements as they had found constitutional standing lacking.

This case was a victory for animal advocates in the area of standing under the APA and the AWA. The aesthetic injury recognized in the earlier cases cited by the dissent involved enforcement of the ESA, the law designed to protect species from extinction. Thus, the interest to human plaintiffs involves being able to view threatened species in the wild. In contrast, the AWA is designed to protect animals from inhumane treatment. Consequently, the human plaintiffs' interest is to witness animals free from inhumane treatment, which this court recognized. Moreover, by holding that amended regulations would in fact remedy the plaintiff's interest here, although subjective and not necessarily addressed by a new regulation, the court provides greater opportunity for private plaintiffs to enforce the AWA. It is important to note in comparing this case with the prior AWA standing cases that the context for enforcement here is the exhibition of animals, not the research facility and their use of animals.

Comparison to the United Kingdom

Research facilities in the United Kingdom that use live animals primarily are regulated by the Animals (Scientific Procedures) Act 1986.[82] The animals protected under the Act are "any living vertebrate and any invertebrate of the species Octopus vulgaris from the stage of its development when it becomes capable of feeling." Thus, birds, mice, and rats are covered. Further, an animal is living "until the permanent cessation of circulation or the destruction of the brain."[83]

In the United Kingdom, each researcher must obtain a "personal licence" and demonstrate that the person has the "appropriate education, training and … experience for the purpose of competently handling the protected animals, applying the specified regulated procedures to the specified classes of animal, and taking responsibility thereafter for the welfare of the animals."[84] A regulated procedure is defined as "any experimental or other scientific procedure applied to a protected animal which may have the effect of causing that animal pain, suffering, distress or lasting harm."[85]

Each project also must be licensed. According to an expert, it is not simple to obtain a project license. He explains that the researcher must in fact justify the procedure and prove the case rather than merely fill out a questionnaire. The application process is governed by Inspectors who are "experts in some areas of research and have access to a wide reservoir of expertise in the form of their colleagues versed in other fields of research."[86] The 3Rs are essential to the application for and granting of the license. The guidelines require strict compliance with the 3Rs as the Secretary of State "shall not grant a project licence unless he is satisfied that the applicant has given adequate consideration to the feasibility of achieving the purpose of the programme ... by means not involving the use of protected animals" and "will review, and may recall and revise, license authorities should suitable replacement, reduction and refinement alternatives become available during the lifetime of a project licence [five years]." Moreover, the Act is designed to fully implement the EU legislation that requires that an "experiment shall not be performed if another scientifically satisfactory method of obtaining the result sought, not entailing the use of an animal ... is reasonably and practicably available."[87]

The licensee must justify the project by evaluating the cost or harm to the protected animal against the benefit to the research. The guidance defines the accepted reasons for granting a project license as: "control of disease, ill-health or abnormality, physiological studies, environmental protection, advancement of biological or behavioural science, education or training, forensic inquiries, or breeding genetically modified animals or harmful mutants."[88] Further, the "likely benefit is primarily derived from the utility of the data or product to result from the programme of work, rather that the importance of the general area of study," and the Secretary "must be of the opinion that the programme of work is likely to meet its stated objectives and that the scientific quality of the work cannot be further improved."[89] The likely costs or harms are defined as "the adverse welfare effects (pain, suffering, distress, or lasting harm) likely to be experienced by the protected animals used,"[90] and the 3Rs "must be demonstrably implemented in the context of the proposed work, minimising suffering, not simply reducing the number of animals used in the objective. The minimum number of animals of the lowest degree of neurophysiological sensitivity should be used in the mildest protocols required to meet the stated objectives."[91] Additionally, the guidance provides a list of protocol severity limits to help grade the cost as unclassified, mild, moderate, and substantial, with each level specifically defined.[92]

Finally, every establishment involved with live research animals – the research, breeding, and supplying establishments – must have a certificate of designation.[93] To obtain a certificate the establishment must designate a veterinary surgeon and animal care and welfare officer to participate in a local ethical review process and must comply with the "codes of practice" that define standards for the animals' housing, care and humane killing.

In comparing the Act of the United Kingdom governing research animals with the United States AWA, it appears that the use of animals in research is more limited and the animals who are used are better protected in the United Kingdom. First, all animals used in research are covered under the Act while less than 10 percent of the animals used for research are covered under the AWA. Second, the United Kingdom licenses not only the facility, but each researcher and each project, arguably providing greater oversight than under the AWA. Moreover, although both countries seek to implement the 3Rs, enforcement in the United Kingdom is likely much stricter since the entity reviewing the project license, the Inspector, is independent of the research institution, unlike the IACUC in the United States.

Animals used for breeding and sold as pets

The AWA governs breeders and sellers of pets if they fall within the definition of dealer. As you will recall, "retail pet stores" are excluded from the definition, as are dealers whose business is arguably de minimus.[94] The agency defines a retail pet store to include anyone who sells directly to the public. [95] This exemption excludes a large number of entities. In fact, many facilities, no matter how large, sell directly to the public either in person or via the internet so as to avoid regulation under the AWA.[96] The regulations promulgated under the AWA, as discussed above, set minimal standards for the feeding, watering, vetting, and physical enclosures of the animals being bred or held for sale as "pets."

Views on the ethics of breeding animals are varied. Many humans accept the breeding of companion animals as proper, while others believe that animals should never be bred because it promotes the view of them as property for the exclusive purpose of serving human interests. Some believe that breeding should be prohibited, at least for a period of time, because so many companion animals, mostly dogs and cats, are homeless and killed in shelters every week. Thus, the public should adopt homeless animals from shelters rather than purchase them from breeders. For those who accept breeding as legitimate, they diverge as to the regulation that is appropriate. Some do not believe any regulation should be imposed, while others understand that breeding must be regulated to protect the animals from inhumane treatment and/or disease or other medical issues that may result from improper breeding methods; if not to protect the animals themselves, then at least to protect the human consumer. Independent of one's position, most people would agree that certain so-called breeders, more commonly referred to as "puppy mills," are deplorable and should be outlawed.

A "puppy mill" is a large-scale commercial dog-breeding operation that emphasizes profit over the well-being of the dogs in their care. These enterprises do not consider genetic quality or integrity when breeding dogs and over-breed female dogs until they are physically depleted. At that time, when they are no longer useful to the breeder, they are killed. The dogs are kept in overcrowded and unsanitary conditions, without proper food, water, shelter, or veterinary care. As a result, many of the puppies are prone to congenital and hereditary conditions that are unhealthy and threaten their life.[97]

Many puppy mills are not regulated under the AWA because they are classified as "retail pet stores." Moreover, even if they are subject to the AWA, the standards do not sufficiently protect the animals' welfare. Thus, some states have begun banning puppy mills and imposing criminal sanctions on those who violate the law. As of 2009, Virginia arguably has one of the most progressive puppy mill laws in the United States. The law contains two critical elements: (1) it limits the number of breeding dogs over the age of one year to 50, and (2) requires that female dogs be bred only between the ages of 18 months and eight years and be certified annually by a licensed veterinarian.[98] Moreover, the law states: "It shall be the duty of each attorney for the Commonwealth to enforce this article."[99] This is an unusual provision and appears redundant because prosecutors are required to enforce the criminal laws. However, prosecutors often have little time, resources, or incentive to enforce the laws protecting animals. This provision emphasizes that protecting dogs from inhumane treatment in breeding facilities is of paramount importance to the citizens of the state of Virginia.

Animals used for entertainment

Defining "exhibitor"

Let's now turn to the regulation of animals used for entertainment or exhibition under the AWA. The AWA defines "exhibitor" as any person (public or private) exhibiting any animals, which were purchased in commerce or the intended distribution of which affects commerce, or will affect commerce, to the public for compensation, as determined by the Secretary, and such term includes carnivals, circuses, and zoos exhibiting such animals whether operated for profit or not; but such term excludes retail pet stores, organizations sponsoring and all persons participating in State and country fairs, livestock shows, rodeos, purebred dog and cat shows, and any other fairs or exhibitions intended to advance agricultural arts and sciences, as may be determined by the Secretary.[100]

It is clear from this definition that not every animal "exhibited" is protected, given the broad exclusions. Exhibitor "*includes* carnivals, circuses, and zoos." What other "exhibition" is "included?" A court addressed this question in *Haviland v. Butz*, when the owner of a traveling dog and pony show brought suit to challenge his inclusion under the AWA as an "exhibitor."[101] The agency regulations implementing the definition of exhibitor under the AWA used the same terminology as the statute but included the words "animal acts" between circuses and zoos. Haviland argued that Congress did not intend such a broad sweep. The court disagreed, noting that the House Committee on Agriculture had referred to "circuses, zoos, carnivals and road shows."[102] The court stated that while there is no explanation for why "road shows" was not included in the final statute, apparently they were contemplated by the Committee. Since the statute indicates that the list is not necessarily complete, the agency's inclusion of "animal acts" is not in violation of the statute.

Haviland then argued that by including his dog and pony show but excluding rodeos Congress violated his right to equal protection, as there was no rational basis to distinguish between the two. The history of the bill showed that the original definition stated "including but not limited to zoos and circuses."[103] Later, without explanation, rodeos were expressly excluded despite evidence that animals are abused in such environments. The court, in trying to find a reasonable explanation for the exclusion, stated that some members of Congress expressed a general concern for the cost of administrating the AWA. Congress excluded retail pet stores, arguably given the number of them in the country, and thus likely excluded rodeos for the same reason.[104] The court further noted that carnivals and circuses travel and often include dog and pony shows. Therefore, Haviland's dog and pony show is more similar to the expressly included acts than to the excluded ones, both in their performances, use of animals, and travel.

The court further stated that it is not unconstitutional for Congress to address some but not all problems related to the inhumane treatment of animals, even those that may appear similar. "Evils in the same field may be of different dimensions and proportions, requiring different remedies; reform may take one step at a time; addressing itself to the phase of the problem which seems most acute to the legislative mind."[105] Moreover, the court noted that evolution of the AWA has been slow, with Congress choosing a "cautious approach ... as the national interest seemed to warrant."[106] Thus, including Haviland's dog and pony show as an "exhibitor" while excluding rodeos did not violate Haviland's constitutional rights.

Arguably the judge arrived at the correct decision. Nevertheless, the rationale behind the excluded acts is not fully explained. Is it that Congress was primarily concerned about the humane transport of animals and the show was included because it traveled? Was Congress more concerned about the use of captive wild animals for entertainment and, although Haviland's act used only domestic animals, circuses and zoos often use wild animals while rodeos do not? Finally, politics and societal values likely played a large role in Congress' decision. Rodeos, in particular, are popular in the United States, and likely have lobbyists representing their interests on the hill. In fact, a "humane rodeo" is an oxymoron of sorts. It likely would be difficult to regulate and impose humane standards on rodeos without outlawing them – the only solution consistent with protecting the animals' inherent interests independent of their owners or the viewing public.

Regulating exhibitors

For entities that are covered as exhibitors, the Code of Federal Regulations (CFR) contains specifications for the humane care, treatment, handling and transport of various animals, as described above, including wild, captive animals. Specifically, the regulations require knowledgeable handlers of wild or exotic animals who must handle the animals in a manner that "does not cause trauma, overheating, excessive cooling, behavioral stress, physical harm, or unnecessary discomfort."[107] Short-term withholding of food and water is allowed so long as the animals receive their daily nutritional requirements. Performing animals must be given periodic rest periods at least equal to the time for one performance. Precautions must be taken to minimize the risk of harm to the animal and the public during exhibition and drugs may not be used to allow for public handling of the animals. Note that these provisions allow for much discretion in interpretation. On the one hand, this is necessary for the regulation to accommodate a variety of exhibits; on the other hand, it does not guarantee that exhibition animals will be free from harm.

In comparison, in the United Kingdom, the Zoo Licensing Act, administered by the Zoo Forum, regulates zoos and circuses.[108] In October 2007, a report by the Circus Working Group, "Wild Animals in Travelling Circuses,"[109] was published. The Circus Working Group, comprised of representatives from humane organizations and circuses, was created by the United Kingdom after the Animal Welfare Act of 2006 was enacted. In 2002, the Department for Environment, Food and Rural Affairs (DEFRA) had issued a circular which stated:

> It believes that all captive animals should enjoy the same minimum welfare standards, aimed at ensuring a quality of life as good as can *reasonably* be achieved in the type of regime in which they are held. They should be held in accommodation, which is suitable in every key respect; adequately fed and watered; provided with veterinary care as necessary; and not subjected to *unnecessary* suffering. Wherever *practicable*, standards should go beyond that – for example, to provide a rich and stimulating environment.[110]

This statement, while arguably designed to promote the welfare of captive wild animals, leaves room for much discretion. In fact, some officials and others believed that allowing traveling circuses to use wild animals was not consistent with meeting their welfare needs and were considering a ban of certain wild animals. At the time, 47 wild animals were in traveling circuses in the United Kingdom. The report concluded that because there was insufficient scientific evidence to prove that the animals' welfare in traveling circuses was any better or worse than in other captive environments, there was no scientific basis for a ban. Moreover, the report repeatedly explained that the working group was to base their recommendations on scientific evidence only, not politics or ethical or moral judgments since those would be outside the scope of their authority and subject to legal challenge. The report recommended that circuses be regulated to provide more specific guidance than that provided under the AWA which imposes a duty on the animals' keepers to take such steps as are reasonable in all the circumstances to ensure that the needs of their animals are met to the extent required by good practice.[111] They proposed the following language based on provisions of the Zoo Licensing Act. Traveling circuses would be required to:

1. Accommodate their animals under conditions which aim to satisfy the biological requirement of the species to which they belong including –
   i. Providing each animal with an environment well adapted to meet the physical, psychological and social needs of the species to which it belongs; and
   ii. Providing a high standard of animal husbandry with a developed programme of preventative and curative veterinary care and nutrition.
2. Prevent the escape of animals and putting in place measures to be taken in the event of any escape or unauthorized release of animals.
3. Prevent the intrusion of pests and vermin into the circus premises.
4. Keep up-to-date records of the circus's collection ...[112]

Under the Zoo Act, public notice of a license request must be given prior to its grant. Zoos are subject to regular inspection, but at least two are mandatory during the four-year license period. The regulations require that animals be handled by trained staff, who may not smoke near the animals or their food. The display requirements mandate that they have sufficient space and "furniture" within the cages to meet the psychological needs of the animals and that predator and prey species not be within eyesight of one another to avoid undue stress on the animals.

In December 2009, DEFRA published a consultation on the use of wild animals in traveling circuses in England[113] and sought input from the public. Three options were considered: a complete ban, voluntary self-regulation, and compulsory statutory regulation. The topics addressed by the consultation included the financial impact of a complete ban and the re-homing of the 38 wild animals currently in traveling circuses, the nature and enforcement of possible regulations, and ways of raising the standard of welfare of these animals. On March 25, 2010, Jim Fitzpatrick, MP, Minister of State for Food, Farming and Environment, published a letter thanking all who contributed and suggested that a ban may be pursued based on the comments received.[114]

In comparison, New Zealand has fairly comprehensive legislation that promotes the welfare of the animals on exhibit expressly.[115] The registration process for facilities requires an animal collection plan and a contingency plan outlining the fate of the animals should the facility fail. Registration and inspection are both annual. A veterinarian reviews each license application for viability and may deny a facility to keep one or more species. The standards are strict. They not only require a high level of hygiene, but specifications for each species govern the following categories: (1) behavioral requirements of individual (that is, swimming, climbing, grooming, territoriality); (2) behavioral requirements of social groups (that is, size/sex ratios, seasonal changes, hierarchies, compatibilities, need to escape conflict); (3) physical requirements (exercise, shelter, individual cover, territories, ventilation); (4) psychological requirements (that is, intellect, adaptability, timidity, aggressiveness); (5) reproductive requirements (reproductive control must be incorporated); and (6) zoographic requirements (that is, expected life span, rate of population increase).

The approaches of the three countries to regulate the care of wild animals when used for exhibition purposes are similar. Arguably, no amount of regulation suffices to provide a captive wild animal the life they were intended to lead – one in the wild. The use of wild animals in circuses, especially traveling circuses, is difficult to justify. It is interesting that the Circus Working Group found insufficient evidence to suggest that a captive wild animal's welfare is not necessarily any worse when living in a traveling circus than in any other captive environment. Although the report reiterated the theme that their findings were to be based solely on scientific evidence and not politics or ethical judgments, it would appear that politics may have played some role in the final analysis. However, if the law is to allow such use, the regulations must be specific, bright-line rules, demanding high standards, with frequent review and strict enforcement and public reporting requirements. The detailed categories as provided under New Zealand law are necessary along with specific criteria to meet each requirement. Moreover, the enforcement must be conducted by independent authorities. The better solution may be a complete ban on their use, as the United Kingdom is currently contemplating.

Ethics of exhibiting animals

The use of animals, domestic or wild, for exhibition purposes is ethically controversial. Domestic animals arguably may be more suitable to exhibition than wild animals given their nature and temperament. However, depending upon the individual circumstances they may still suffer harm from the confinement and stress of public scrutiny. However, as discussed above, in the United States, many exhibitions of domestic animals, like rodeos, fairs, and horse and greyhound racing, are not covered within the regulations and thus there are no legal protections for these animals.

Using captive wildlife for exhibition is of special concern. On the one hand, if the animals are well cared for and the exhibition itself does not cause physical or emotional stress to the animals, one could argue that the benefits outweigh the minimal harm to the animals. Such use provides humans an opportunity to see animals they otherwise would not be able to see and enhances our views of their lives. This increases our respect for and appreciation of their existence while also providing entertainment. Moreover, the captive wild animals are often bred to help promote the propagation of the species.

However, the reality is that the cost to the animals is great. Wild animals by definition are no longer wild but captive and this change in their environment itself causes substantial harm. Moreover, in exhibitions such as circuses (as compared to zoos, for example) training techniques typically involve conduct that injures the animals. Furthermore, the benefits indicated above are not realized. Seeing animals in captivity, performing in a circus and even being held in most zoos, does little to enhance our understanding or appreciation for the animal and their species. The public does not see the animals engaging in their natural behavior – searching for food, caring for and protecting their young, or socially interacting with others of their species. Instead, the public often sees animals that are bored and/or neurotic. Finally, to the extent we must capture a wild animal in an effort to propagate the species because we have endangered them, the animals need not be exhibited.

The 2003 case of *Born Free USA* v. *Norton*[116] presented the ethical issue of keeping wild animals captive. In the late 1980s Swaziland had imported African elephants from a national park in South Africa to reintroduce the species to their Kingdom. They lived in reserves operated by Ted Reilly, the government official for managing the threatened and endangered big game. By 2002 the reserves had grown to 30 adults and six calves and Reilly had become concerned about the impacts on biodiversity. The elephants can deplete vegetation, damage trees that are the home to birds, and compete with the black rhinoceroses, even more endangered than the elephants, for resources. Reilly determined to remove eleven elephants. Two zoos in the United States had made arrangements to import the elephants both for display and to revive their captive breeding programs. The zoos hoped that the established social bonds of these elephants would enhance their breeding and would increase the genetic diversity of the captive elephant population in the United States. Under the Convention on International Trade in Endangered Species (CITES), the parties were required to obtain import and export permits from the United States Fish and Wildlife Service. The permits were approved.

The zoos were preparing to transport the elephants when, on July 10, 2003, Born Free USA brought suit challenging the grant of the permits, under CITES, the ESA, and the National Environmental Policy Act (NEPA), alleging injury to the eleven elephants and the Swaziland elephant herd that would remain if the eleven elephants were imported to the zoos. The plaintiffs sought a preliminary injunction to prevent their transport until the case was decided. The eleven elephants were being kept in a corral in Swaziland and had been there for several months awaiting export. Reilly stated that he could not hold the elephants past mid-August. If the elephants were not transported they would be killed. Ultimately, the court had to determine what was worse for the elephants, death or transport to the zoos.

Of interest here were the difficult competing interests and harms presented to the court in deciding whether to issue the preliminary injunction. The court explained that its principal consideration was what would happen to the elephants if the injunction was granted. The elephants were already separated from the herd, in the corral in Swaziland. This created "pathogenic threats to [the elephants'] health;" placed "unnecessary strain on tractor haulage and human resources, maintenance of water installations (during the persisting drought), and other park infrastructure;" and required 24-hour armed security to guard the elephants, which diverted their efforts to protect rhinos and other elephants from poachers.[117] The elephants would be killed if the preliminary injunction was issued, as Reilly would not retain them past mid-August. Although plaintiffs claimed they had found suitable homes for the elephants, Reilly stated he would not pursue other alternatives – the elephants either went to the zoos or they were killed. The court noted that while killing the elephants may seem offensive, it had been previously employed to reduce populations in southern Africa. The plaintiffs explained that their interests would be irreparably harmed if the court allowed them to be imported to the zoos.

If the court denied the injunction, it was unclear what would happen to the elephants if the plaintiffs ultimately proved their case. The FWS could attempt to export the elephants back to Swaziland, but it was implausible to believe that Swaziland would accept them. Moreover, CITES would make it difficult to export them to any other country if the court had found that the initial importation had been invalid. Thus, the zoos may end up keeping the elephants anyway. At a hearing on the preliminary injunction, the plaintiffs admitted that given the choice, they would rather see the elephants dead than in a zoo.[118]

The zoos had an interest in importing the elephants for their breeding programs, and therefore killing the elephants would substantially harm their interests. However, the court noted that the zoos could probably import elephants from other countries to fill their needs. As for the federal government and public interests, the court found them in equipoise. The FWS was a detached government body that did not have a direct interest in these elephants. The public, on the one hand, according to the court, would benefit from being able to view the elephants in the zoo and this would promote and encourage conservation efforts. On the other hand, the public may consider zoos improper places to keep wild animals and thus might believe that death is a better alternative for these elephants. The court acknowledged that saving the elephants from death may be publicly more acceptable, but there was no clear evidence which view the public would adopt.[119]

The court ultimately denied the injunction allowing the transport of the elephants to the zoos based primarily on its finding that the plaintiffs were not likely to succeed on the merits and the zoos' interests would be harmed significantly if it were granted. Of course, it is unknown how the court would have ruled had it believed the plaintiffs would likely succeed on the merits. As discussed above, if erroneously denied, as a practical matter, the elephants may have remained in the zoos anyway. The situation presented was further complicated by the fact that Reilly had already taken the elephants from the herd, well before they were to be exported, and the United States judge had no authority to enjoin Reilly, in Swaziland, from killing the elephants or finding another suitable location during the course of litigation had he granted the preliminary injunction.

A case study: seeking to protect Asian elephants in the Ringling Brothers' Circus

The use of wild animals, especially elephants, in circuses is very controversial. Ringling Brothers and Barnum & Bailey Circus, operated by Feld Entertainment, Inc. (FEI) is allegedly "The Greatest Show on Earth." They use Asian elephants as performers in their circus. These elephants are highly intelligent and social animals. In a circus environment they are chained on hard surfaces for most of their lives, routinely struck with bull hooks to perform, and transported on trains in chains for days at a time. In addition, the young are prematurely separated from their mothers and the adults are forced to perform unnatural acts solely for the pleasure of the human audience and the circus owner's profit. Nevertheless, the United States does not prohibit circuses from using these animals. Instead, the law regulates the conditions under which they are held, although USDA enforcement has been ineffective. USDA investigators routinely documented serious violations of the AWA in connection with FEI's treatment of the Asian elephants, yet they failed to impose serious fines on FEI. As a result, FEI had merely included the relatively minor penalties, if any, as the cost of doing business and continued to mistreat the elephants.

The American Society for the Prevention of Cruelty to Animals (ASPCA) and other humane organizations, frustrated with USDA's ineffective enforcement, wanted to legally challenge FEI's alleged abuse of these magnificent creatures but they could not use the AWA because they had no standing. What options did they have, if any? The lawyers had an idea.[120] The Asian elephant is listed as endangered under the ESA and, as such, may not be "taken" under the ESA unless the entity has a permit. Permits are granted only if the "take" is "for scientific purposes or to enhance the propagation or survival of the affected species."[121] FEI had a Captive-Bred Wildlife (CBW) permit[122] for only 21 of the 28 Asian elephants. The permit requires that "live wildlife ... be maintained under humane and healthful conditions,"[123] and that the permit holder comply "with all applicable laws and regulations governing the permitted activity."[124] Among the "applicable laws and regulations" is the AWA regulation that states "physical abuse shall not be used to train, work or otherwise handle animals," that "handling of all animals shall be done ... in a manner that does not cause trauma ... behavioral stress, physical harm, or unnecessary discomfort," and that "young or immature animals shall not be exposed to rough or excessive public handling."[125]

Since the ESA allows for a private right of action, the Asian elephants are endangered, and FEI had no permit for seven of the elephants and allegedly was in violation of the CBW permit they had covering the 21 elephants, the plaintiffs filed suit against FEI for violations of the ESA. They alleged that the treatment of the Asian elephants by FEI "harassed,"[126] "harmed"[127] and "wounded" these creatures and that FEI was guilty of "taking" them in violation of Section 9 of the ESA without a permit and/or in violation of the permit requirements that they be maintained under humane and healthful conditions and in compliance with AWA regulations. The defendant, FEI, denied the allegations of abuse and raised a number of defenses to defeat the suit,[128] including standing and failure to state a claim. Nevertheless, the lawsuit proceeded and for several years the parties engaged in, and argued over, discovery.[129]

In 2007, FEI filed a motion for summary judgment to dismiss the suit involving the 21 CBW permit elephants on the ground that plaintiffs lacked the jurisdiction to challenge FEI's compliance with the CBW permit issued by the FWS.[130] Although plaintiffs alleged that the CBW permit is supposed to "enhance the propagation of the species," and FEI's treatment of these elephants directly contradicts such purpose, the court never addressed the merits. Instead, the court agreed with FEI that the plaintiffs had no private right of action to enforce compliance with the permit. The court explained that Section 11 of the ESA contemplates a broad enforcement scheme by various entities. The citizen provision suit of 11(g) allows a private citizen to enjoin a violation of the ESA or its regulations. However, only 11(a), (b) and (e) specifically reference *permits* issued under the ESA and only provide for enforcement by government actors. Thus, the clear intent of Section 11(g) is to disallow private citizens to bring suits to enforce compliance with issued permits.

The case proceeded, but only against the seven elephants not covered under the CBW permit – Jewell, Lutzi, Mysore, Susan, Zina, Nicole and Karen. Two major legal issues that the parties debated was whether the take provision of Section 9 applies to captive wildlife and what the appropriate scope of relief would be if the plaintiffs won.

Does the "take" provision of the ESA apply to captive wildlife?

Can an entity "take" under the ESA an endangered animal who has already been removed from the wild? Defendants argued that when the word "take" is read in context with the explanatory terms "harass, harm, pursue, hunt, shoot, wound, trap, capture, or collect," the intent to limit application of a "take" to animals in the wild becomes clear. They stated:

> it makes little sense that animals in captivity would be pursued, hunted, shot, killed, trapped, captured, or collected. And if "wound" or "harm" apply to animals in captivity, then all endangered species who receive veterinary care or have their food and shelter provided are arguably being taken … Simply put, if "take" applies to captive species, then anyone who holds an endangered species in captivity, whether a circus, zoo, or wildlife sanctuary, is violating the "taking" prohibition.[131]

FEI further argued that even if it does apply, only "harass" would logically apply, and FEI's treatment is not harassment. First, the term "harm" cannot apply to captive animals because it includes "habitat modification" and "wound" means any penetration of the animal's hide. Thus, if applicable, all animal husbandry practices and veterinary care itself would "harm" and "wound" the animals. Moreover, FWS amended the definition of "harass" to exclude "generally accepted … animal husbandry practices that meet or exceed the minimum standards for facilities and care under the [AWA]."[132] Thus, if "harm" and "wound" applied to captive animals, it would render the definition of "harass" meaningless. This is so because every captive animal would be "taken" by virtue of being cared for, even if the care exceeded AWA standards, because "animal husbandry practices" would still be considered harm although not harassment. Second, since the use of guides and tethers are generally accepted husbandry tools, and the USDA has concluded that "striking an elephant with a guide and creating a bloody wound is *not* a violation of the AWA," FEI's handling of the elephants complies with the AWA as administered by the USDA and is thus not a "take."[133]

The plaintiffs argued that the take provision absolutely applies to captive animals and that FEI's treatment of the elephants harms, wounds, and harasses them. First, the plain language of the statute broadly states: "with respect to any endangered species … it is unlawful for any person … to … take any such species."[134] The statute does not limit the species in any manner, but rather emphasizes "any." Second, the grandfather clause that exempts listed species held in captivity at the time of enactment of the ESA was amended in 1982 to remove the "take" provision as an exemption.[135] Third, the FWS has stated that the statute applies to wild and captive populations of listed species. In fact, given the FWS promulgated captive-bred wildlife regulations and these regulations have special provisions for captive members of certain species, the statute must apply to captive wildlife otherwise these provisions would be irrelevant. Fourth, including all endangered species, wild and captive, is most consistent with the underlying purpose of the ESA – to provide comprehensive protection for all endangered and threatened species.[136]

Further, the plaintiffs agreed with the defendant's argument that such an interpretation would mean that anyone who has a captive endangered species is violating the "take" provision *unless they have a permit*. This is precisely what the ESA requires, a permit to "take" an endangered species and only "for scientific purposes or to enhance the propagation or survival of the affected species."[137] Moreover, zoos and wildlife sanctuaries do not chain their elephants for hours on end on hard surfaces, transport them around the country in trains for weeks at a time, nor beat them with bull hooks. Thus, they are not engaging in the explicit conduct that is harassing, harming, and wounding the elephants.

Finally, all three terms – "harm," "wound," and "harass" – apply to captive animals and does not render "harass" meaningless. This is so because "harm" and "wound" involve conduct that injures an animal not conduct that benefits the animal. Thus, even if veterinary care may involve penetration of the animal's hide, it would not be deemed a "wound" because it would not be injurious to the animal.[138] Moreover, the exemption clause including normal animal husbandry practice was required for the term "harass" precisely because the mere act of holding an animal in captivity significantly disrupts their normal patterns, which is the regulatory definition of "harass."[139] Thus, in order to facilitate breeding of captive species necessary to propagate the species, such exemption is necessary. However, physical mistreatment, like beating an elephant with a bull hook, is not normal husbandry practice and thus constitutes harassment.

What relief is proper should the plaintiffs win?

The plaintiffs requested declaratory and injunctive relief if the court determined that FEI was taking the Asian elephants in violation of Section 9 of the ESA. The declaratory relief is exactly that, a declaration that FEI is violating the ESA with respect to the non-permitted elephants. FEI questioned what such a declaration would mean for the other 21 elephants that they maintain under the CBW permit. Arguably, such conduct would by definition violate the CBW permit and the FWS would be obliged to take action to enforce the permit and end the abuse of those elephants as well.

The injunctive relief is a bit more difficult to define. What they sought was a court order prohibiting FEI from engaging in the "taking" activities, including continual chaining and beating of the elephants with bull hooks, at least until FEI was able to secure a permit from FWS to allow such activities to continue. FEI argued that they would be unable to secure such a permit (which one would hope would be true) and that they then could not maintain the elephants without engaging in such conduct because these activities are required to train and manage elephants in a circus environment. Thus, the elephants would have to be transferred elsewhere to facilities that would not need to engage in such activities (such as zoos or wildlife reserves). This would raise difficult practical issues, such as where would they be transferred and under what terms and conditions.

In 2009, the case went to trial. The trial lasted several weeks, with the plaintiffs presenting extensive evidence of FEI's treatment of the elephants and expert testimony on the effect of this treatment on the elephants both physically and emotionally.[140] After closing arguments, the parties returned to the court twice arguing post-trial motions. On December 30, 2009, the court ruled for the circus.[141] The issues over which the parties argued were never decided since the court dismissed the case on standing grounds. It is difficult and frustrating for animal protection attorneys to be confronted over and over again with courts deciding cases on standing grounds. Moreover, it is unusual for a judge to allow the case to continue through years of discovery, trial, and post-trial motions, only to dismiss the case on standing, an issue often decided at the beginning of the litigation.

This case further demonstrates how important it is for the law to grant animals standing to enforce the laws allegedly designed to protect them. In a statement to the press after hearing of the court's decision,[142] Adam Roberts, executive vice president of API stated: "We are disappointed that the judge dismissed the suit without considering the merits."[143] Defendant Feld Entertainment's chief executive Kenneth Feld said that they "are gratified with today's decision because it is a victory for elephants over those whose radical agenda, if adopted, could lead to the extinction of the species." This comment is quite misleading. First, the decision was not on the merits and thus there was no finding that the elephants are not being subjected to extreme abuse at the hands of the defendant. Thus, for these elephants it is unlikely a victory. Moreover, there is no need to chain these magnificent animals for hours on end, transport them on trains for days at a time, and force them to perform in a circus in order to save them from extinction. Elephants are bred by zoos and wild animal sanctuaries, thus preventing their use for entertainment in circuses clearly would not lead to their extinction.

Animals used for food

Introduction: factory farming

Although "farm animals represent ninety-eight percent of the animals raised and killed"[144] in the United States, with the slaughter of *1 million animals per hour,* virtually *no* laws protect them.[145] Animals used for food are exempt from coverage under the AWA and there are no federal laws that regulate their treatment while on the farm.[146] As a result, animals raised for food are cruelly treated by large factory farming operations. These factories often are referred to as Concentrated Animal Feeding Operations (CAFOs), a special type of Animal Feeding Operation (AFO). The United States Environmental Protection Agency (EPA) defines AFOs as "agricultural operations where animals are kept and raised in confined situations. AFOs congregate animals, feed, manure and urine, dead animals, and production operations on a small land area. Feed is brought to the animals rather than the animals grazing or otherwise seeking feed in pastures, fields, or on rangeland."[147] If the animals are confined for at least 45 days in a twelve-month period, and "there's no grass or other vegetation in the confinement area during the normal growing season," the operation is an AFO.[148] If, in addition, the AFO either confines a large number of animals as defined by the EPA or uses a certain type of pollutant discharge method, the AFO is a CAFO.[149]

Factory farming operations maximize profits to the detriment of the animals' health and welfare by keeping many animals in a small area, resulting in high-density stock. These industries use various physical constraints, including tethering and small cages and pens that prevent the animal from moving freely, artificial methods to promote growth, and antibiotics to offset the spread of disease from the high-density confinement, all to the detriment of the animals' welfare. Moreover, factory farms deplete natural resources, produce high levels of waste that pollute the environment, and have been linked to a variety of human health concerns including illnesses relating to E. coli and salmonella-contaminated meat, and influenza viruses.[150]

The United States government not only allows such factory farms to exist but subsidizes them under the Farm Bill. The Agricultural Investment Act of 1933 was enacted as part of the New Deal following the Great Depression with the purpose "to raise crop prices, combat hunger, conserve soil, provide farm credit and crop insurance, 'and build community infrastructure for rural farming towns.'"[151] The bill provided price supports for 100 different crops and resulted in a 50 percent increase in farm income within three years. Over the years, however, with technological developments applied to food production, larger farms swallowed the smaller farms, and the bill designed to help small farms survive instead morphed into huge government subsidies for a mere five crops, independent of consumer demand. From 1995 through 2006 the U.S. government gave out over $170 billion in agricultural subsidies; almost one-third in subsidized corn.[152] The result is that today almost half of this huge corn supply is fed to animals in factories rather than humans. Americans, in turn, eat much more meat than a few decades ago; on average, 54 percent more than in 1950, with a "national total of 9.5 billion animals raised and butchered each year"[153] to satisfy this demand. Yet the welfare of these animals while in the factory is ignored by the federal government.

Regulating slaughter and transport

There are two federal statutes in the United States that govern animals used for food when not on the farm – the Humane Methods of Slaughter Act (HMSA) enacted in 1958, and the Twenty-Eight Hour Law, enacted in 1877. The Findings and Declaration of Policy of the HMSA make clear that although preventing needless suffering was a motivating factor, benefits to humans were primary. The Findings state:

> The Congress finds that the use of humane methods in the slaughter of livestock prevents needless suffering; results in safer and better working conditions for persons engaged in the slaughtering industry; brings about improvement of products and economies in slaughtering operations; and produces other benefits for producers, processors, and consumers which tend to expedite an orderly flow of livestock and livestock products in interstate and foreign commerce.[154]

Congress lists several benefits, only one of which is a benefit to the animals. The primary benefits are to the workers, slaughtering operations, and consumers. Moreover, the supposed benefit to the animal is to prevent "needless" suffering. Congress does not define "needless" here, but one could argue that, since humans have no "need" to eat the flesh of animals, any suffering by an animal in order to provide food for humans is "needless."

The House Report of the HMSA when amended in 1978 stated:

> While the HMSA had its genesis in concern for the humane treatment of animals, meat packers and processors soon found an economic incentive to adopt humane methods of slaughter and of handling in connection with slaughter. Humane methods proved to be more efficient and less hazardous to plant personnel and such methods also eliminated much of the bruising and other damage to meat which had been the occasion of significant financial loss to the industry.[155]
> ...
> Witnesses have testified to having been sickened, physically as well as emotionally, upon learning of cruel abuses to livestock from which food they were eating or had eaten was derived. Some changed diet and now needlessly forgo a valuable source of protein out of revulsion at the inhumane manner in which some livestock are handled. The committee believes that this bears directly on the health of American consumers.[156]

This statement further supports the proposition that the HMSA is intended to benefit the agricultural industry. It is telling that Congress is concerned about the "bruising and damage" to "meat," not the injury to the animal. Furthermore, Congress suggests that the HMSA will alter the choices made by consumers who changed their diet and stopped purchasing meat products after being sickened physically and emotionally from hearing of the cruel abuses to the animals, which, in turn, bears directly on the health of the consumer. It is unclear whether

Congress is concerned about the physical or the emotional health of the consumer. Arguably, if the consumer is emotionally sickened by the inhumanity of the slaughtering operations, their emotional health may improve if conditions are improved. Regarding physical health, many chronic diseases have been linked to an over-consumption of animal products and a decrease in consumption of whole plant-based foods. For example, faculty at The Harvard School of Public Health have recently constructed the Healthy Eating Pyramid[157] with whole grains, nuts, seeds, fruits, vegetables, healthy non-animal fats, and tofu as the primary ingredients of a healthy diet. Thus, encouraging consumers to purchase animal products may not benefit their physical health. Perhaps a primary goal was to bring those consumers back to meat products in order to increase the profits of the meat product industry.

Humane slaughter is defined as:

(a) in the case of cattle, calves, horses, mules, sheep, swine, and other livestock, all animals are rendered insensible to pain by a single blow or gunshot or an electrical, chemical or other means that is rapid and effective, before being shackled, hoisted, thrown, cast, or cut; or
(b) by slaughtering in accordance with the ritual requirements of the Jewish faith or any other religious faith that prescribes a method of slaughter whereby the animal suffers loss of consciousness by anemia of the brain caused by the simultaneous and instantaneous severance of the carotid arteries with a sharp instrument and handling in connection with such slaughtering.[158]

The critical aspect of humane slaughter is to use a method that itself does not cause pain (for example, a gunshot) to render the animal insensitive to pain before it is processed (such as shackled, hoisted, and so on). Recall from Chapter 2, however, that the second method fails to render the animal insensitive to pain prior to being shackled and hoisted and thus arguably is inhumane.

The Twenty-Eight Hour Law governs the transport of animals, including those raised for food, in interstate commerce via "rail carrier, express carrier, or common carrier (except by air or water)."[159] The law states that the animals may not be confined "in a vehicle or vessel for more than 28 consecutive hours without unloading ... [for at least five consecutive hours[160]] for feeding, water and rest."[161] However, "sheep may be confined for an additional 8 consecutive hours ... when the 28-hour period ... ends at night,"[162] and other animals may be confined for longer periods in the event of "accidental or unavoidable causes," or for 36 consecutive hours at the written request of the owner.[163] These exceptions provide much opportunity to allow transport of animals for over 28 consecutive hours.

The USDA administers and enforces these laws. Pursuant to the HMSA, regulations prescribe standards and specifications for livestock pens, driveways and ramps, the handling of livestock, the tagging of equipment, and the various methods available for slaughter, including chemical, carbon dioxide; mechanical, captive bolt; mechanical, gunshot; and electrical, stunning or slaughtering with electric current.[164] The regulations implementing the Twenty-Eight Hour law provide guidelines for the handling and feeding of animals during transport by species of animal.[165]

In practice, these laws are ineffective. During the first 130 years of enforcement of the Twenty-Eight Hour law, no animals transported by truck, a primary means of transport in recent decades, were protected because the USDA did not interpret "vehicle" to include trucks! In 2006, in response to a petition challenging that exclusion, the USDA finally agreed that the plain meaning of the term "vehicle" includes trucks.[166] Moreover, the law is rarely, if ever, enforced and the penalties are meaningless: $100–500 for violation per shipment (not per animal). Most importantly, however, 95 percent of all animals slaughtered for food are not protected under either law because fish, rabbits, and birds[167] are not covered.[168]

A case study: statutory and regulatory interpretation excluding birds from the HMSA

In 2005, plaintiffs representing poultry eaters and poultry slaughterhouse workers (not the birds or the birds' interests) brought suit challenging the USDA's interpretation of "livestock" under the HMSA excluding all birds.[169] This case presented an exhaustive exercise in statutory interpretation. The federal trial court, in detail, discussed every factor relevant to the interpretation of a statute: the express language, including resort to traditional canons of construction, legislative history and intent, and policy arguments to decide that fish and birds are excluded from the term "livestock" under the HMSA.

First, the court interpreted the statutory language. Recall that the HMSA states that the following animals are covered: "cattle, calves, horses, mules, sheep, swine, and other livestock."[170] Since "livestock" is not defined in the statute, plaintiffs argued that the standard dictionary definition of "livestock" controls. Webster's defines "livestock" as "domestic animals used or raised on a farm – especially those kept for profit."[171] Defendants, using a different dictionary, argued that the proper definition is "horses, cattle, sheep and other useful animals kept or raised on a farm or ranch."[172] The court, after considering both arguments, stated that the definition could be "limited to a narrow group of quadrupeds like cattle and other bovine creatures or alternatively, it could be all-encompassing." Moreover, "there do not seem to be any dictionary definitions from the 1950s that explicitly include poultry as livestock," which is relevant because dictionary definitions change over time.[173] Furthermore, the USDA, courts, and Congress have been inconsistent in their use of the term "livestock;" sometimes including poultry, and other times not.[174] Thus, the court decided the term "livestock" is ambiguous.[175]

Since the statutory language is ambiguous, the court turned to a review of the legislative history. The court noted that the 1907 enactment of the Federal Meat Inspection Act (FMIA), regulating the production and distribution of meat, covered "cattle, sheep, swine, goats, horses, mules, or other equines."[176] In 1957, one year before enacting the HMSA, Congress enacted the Poultry Products Inspection Act (PPIA) to comprehensively regulate poultry products in order to protect public health from adulterated food, but did not treat methods of slaughter.[177] The HMSA has been amended several times since, and poultry has never been included expressly. Congressional debate surrounding the HMSA was ambiguous, with some members suggesting poultry was included and others suggesting it was not. Moreover, often when Congress intends to include poultry when discussing meat or livestock, they do so expressly. Thus, the legislative history also is ambiguous.

Thus, the court referred to the traditional canons of *noscitur a sociis* and *ejusdem generis* because the term "other livestock" follows a list of enumerated animals.[178] The former canon provides that when a word is ambiguous, its meaning may be determined by reference to the rest of the statute. The latter provides that where general words follow an enumeration of specific items, the general words are read as applying to other items akin to those specifically enumerated. The court concluded that these canons suggest that Congress intended to limit livestock to other quadrupeds. Plaintiffs tried to persuade the court that the phrase "and other livestock" was to include a separate category from the other listed terms. However, although the list clearly was meant to be partial, goats – animals who are also quadrupeds – are livestock but not listed, suggesting that the impartial list was that of quadrupeds. Thus, the court held the USDA's interpretation of "livestock" as excluding poultry was proper.

Interestingly, perhaps the most persuasive policy argument for including poultry cut against the plaintiffs. The plaintiffs explained that

98 percent of all animals slaughtered are poultry and thus they must be included to effect humane slaughter, the intent of the HMSA. However, the court noted that if poultry represents such a large percentage of animals slaughtered, Congress would have expressly listed poultry had they intended poultry to be covered.[179] The end result appears to confirm that Congress' primary objective in enacting the HMSA was to benefit humans rather than to protect the majority of animals killed for food.

Some hopeful developments at the state level

There have been recent developments at the state level that give some limited hope to animals used for food. Animal advocacy organizations have been somewhat successful in raising public awareness of the cruel factory farming practices. These practices include: battery cages – small wire cages stacked in tiers and lined up in rows inside huge warehouses with four hens in one cage measuring just 16 inches wide; sow gestation crates – two-foot by four-foot crates barely larger than the sow's body; and the forced feeding of ducks to produce foie gras.[180] The groups then generate sufficient popular support to enact laws through the initiative process to prohibit these practices. In 2002, Florida voters passed a constitutional amendment entitled "Limiting Cruel and Inhumane Confinement of Pigs During Pregnancy" that banned gestation crates effective 2008. In 2006, Arizona voters passed the Humane Treatment of Farm Animal Act banning veal crates and gestation crates effective 2012. In 2007, the Oregon legislature banned gestation crates effective 2013; and in 2008, California voters passed the Prevention of Farm Animal Cruelty Act banning gestation crates, veal crates, and battery cages effective 2015. In 2009, the Maine legislature banned veal and gestation crates effective 2011, and Michigan banned gestation crates effective 2012 and veal crates and battery cages effective 2019.[181] Although there are exemptions in these laws for certain practices (typically practices not involving food production directly) and there are fairly extensive phase-out periods (some, ten years), the trend appears to offer a glimmer of hope for improving the welfare of animals used for food in the United States.

However, the hope may be short-lived. During the same 2009 election where the animals won in Maine and Michigan, the animals of Ohio suffered what all humane organizations considered a defeat[182] and which may signal a reversal of this trend in the states to improve the welfare of animals used for food. Ohio voters passed "the Ohio Livestock Care Standards Amendment, which would create a 13-member board consisting of farmers, experts in food safety, the state's veterinarian, and consumer groups that would be tasked with establishing animal welfare standards in the state."[183] Although packaged as a pro-animal welfare amendment, the amendment was introduced by the agricultural community in reaction to the various states that had recently banned confinement systems. The amendment states that the Livestock Care Standards Board must "endeavor to maintain": (1) "food safety," (2) "locally grown and raised food," and (3) the "protect[ion of] Ohio farms and families."[184] Moreover, the alleged purpose of the Board is "to establish standards governing the care and well-being of livestock and poultry in [the] state but in carrying out this purpose, the board is not required to consider the health and happiness of the animals, and need only attend to worthy, but unrelated, impacts on human health, human farms, and human families."[185] The apparent purpose was to appease voters by leading them to believe that a state board will establish standards to protect the welfare of animals used for food, when in reality the board will be promoting the interests of the farming industry.

In fact, in the case of foie gras, we have already seen how short-lived a victory can be. In 2006, Chicago, Illinois, became the first major city in the United States to ban the sale of foie gras in all of its restaurants.[186] The law survived a constitutional attack by an industry group representing the restaurant owners who claimed the law exceeded the power of the Chicago Council under the Illinois Constitution and violated the Commerce Clause of the Federal Constitution,[187] only to be repealed in 2008 by the Chicago Council under political pressure by restaurateurs.[188]

Thinking outside the battery cage

Promoting the humane treatment of animals used for food directly under the law is quite controversial given the large factory farming enterprises with money and political clout. However, lawyers have begun to use more novel approaches to promote indirectly the humane treatment of animals on the farm through the use of consumer protection and environmental laws. The key, both politically and legally, is to identify *human* interests that are harmed by the CAFOs' treatment of animals used for foods. From a political standpoint, of course, it is easier to persuade government agencies, the courts, and industry to alter conduct in order to protect humans rather than animals. From a legal standpoint, cases can be filed more easily because the humans that are harmed have standing and the claims they are raising are more "traditional."

An example is a campaign to improve the conditions of egg-laying hens in the United States through a variety of legal strategies.[189] First, lawyers target producers who are misleading consumers and file suit for false advertising. The United States has fairly strict laws against false advertising of food products to protect consumers.[190] Thus, when egg producers use deceptive marketing strategies, such as placing the statements "animal-friendly," "certified animal care," or "naturally raised hens" on their cartons, or show images of hens outside on a green pasture or of hens laying eggs in nests when in fact the hens spend their entire lives in battery cages, consumers deceived by these strategies may file suit for false advertising. "Not only is the egg industry cruelly confining hens in cages, it's also deceiving consumers about that abuse."[191]

The remedy is an injunction preventing the defendant from continuing these false representations. The result is not necessarily better conditions for the hens. However, it may result in better conditions because the defendant will likely lose sales if no longer able to misrepresent their production methods. Studies have shown that a majority of American adults find the strict confinement of hens unacceptable, and at least half consider production methods important when purchasing eggs.[192] Thus the defendant may choose to improve conditions for the hens, advertise this to consumers, and thus increase sales.

Second, lawyers work with government agencies to regulate the terms that are used such that consumers understand what the terms in fact mean. For example, the Food Safety and Inspection Service (FSIS) of the USDA regulates the terms "free range" and "free roaming" – "the poultry has been allowed access to the outside"[193] – for certain poultry products, although not for eggs. Note that this definition does not necessarily mean that the birds are free from cruel treatment.

Third, lawyers petition the Food and Drug Administration (FDA), the primary federal agency charged with regulating product labels, to require labels that disclose production methods on animal food products.[194] While the United States has not yet acted to require such disclosure, mandatory labeling on eggs and egg cartons has already been implemented throughout the European Union (EU) and in several states in Australia. All EU eggs and egg cartons must contain a code indicating one of four production schemes: organic, free range, barn, or caged.[195] The goal of these indirect strategies is to use consumer interest and demand and market pressure to persuade producers to change their production methods in order to benefit everyone – themselves, the consumers, and the hens.

Progressive approaches: The United Kingdom and the European Union

In stark contrast to the United States, the United Kingdom has enacted extensive laws protecting animals used for food. In fact, in 2006 when the

United Kingdom rewrote its animal welfare laws it extended protections originally only granted to animals used for food to other domesticated animals. Granted, the laws are far from perfect, but in comparison to the United States they appear quite progressive.

The United Kingdom has been a trendsetter from the beginning.[196] The Brambell Report, drafted in response to a critique of intensive farming practices by Ruth Harrison[197] in her ground-breaking book, *Animal Machines*, was published in 1965. The report addressed the welfare of animals used for food and reported that animals used for food are *sentient*, showing "unmistakable signs of suffering from pain, exhaustion, fright, frustration ... and can experience emotions such as rage, fear, apprehension, frustration and pleasure."[198] The report recommended legislation to protect their welfare. Many of the recommendations were included in legislation and the "Codes of Recommendations for the Welfare of Livestock." The Codes are recommendations and thus not mandatory or directly enforceable. However, noncompliance can be used as evidence for prosecution under the 2006 Animal Welfare Act for causing unnecessary suffering. The Department for Environment, Food and Rural Affairs oversees the welfare of animals used for food (as well as other kept animals except those used in research) and enforces the Animal Welfare Act of 2006.

In addition, the Conventions of the Council of Europe regarding animal welfare and, more recently, EU directives and regulations, provide protections for the welfare of animals. The European Council Conventions relating to animals set forth the following ethical principle: "for his own well-being, man may, and sometimes must, make use of animals, but that he has a moral obligation to ensure, within *reasonable* limits, that the animal's health and welfare is in each case not *unnecessarily* put at risk."[199] In 1979, the European Convention for the Protection of Animals for Slaughter was adopted. This convention controls the conditions in the slaughterhouses and humane stunning methods of slaughter to minimize pain and suffering.[200] In 2004, the European Convention for the Protection of Animals during international transport (revised) was adopted, establishing that animals shall be transported in a way that safeguards their welfare and governs their handling, feed, water, rest and general conditions during transport. The regulation for transport states: "No person shall transport animals or cause animals to be transported in a way likely to cause injury or *undue* suffering to them."[201] Further, the animals must have the "opportunity to rest as appropriate to their species and age, at *suitable* intervals."[202] In 2007, a new regulation established minimum standards for all new vehicles used for the transport of animals over eight hours, and closer monitoring of compliance. Note that while these Conventions are designed to promote animal welfare, the terms italicized above provide for considerable discretion in balancing human benefits against the costs to the animals.

In 2009, the Lisbon Treaty was ratified by all member states of the EU and provides that animals are sentient beings and that member states must take this into consideration whenever legislation is drafted. While the practical and legal effect of this provision is yet to be realized, it is monumental that such a provision is included and signals to all EU citizens that the welfare of animals is important. The EU has adopted several directives regulating the transport, slaughter, and on-farm treatment of animals in the EU, and thus all member states must comply at a minimum with these EU directives. EU directives are binding on the member states as to the result to be achieved, but leave to each member state the method to achieve the result.

The following EU directives regulating on-farm treatment[203] have been enacted to date:

(1) 1998 directive concerning the protection of animals kept for farming purposes ensuring that farmed animals are not caused unnecessary, pain, suffering, or injury, and that their living conditions meet their physiological and ethological needs in accordance with established experience and scientific knowledge.

(2) 1999 directive laying down minimum standards for the protection of laying hens banning unenriched battery cages in 2012 and providing standards for alternative systems and enriched cages.

(3) 2001 directive amending the 1991 directive laying down minimum standards for the protection of pigs requiring, inter alia, accommodations that allow the pigs to rest and get up normally, piglets to be weaned after 28 days of age, and restrictions on mutilation procedures.

(4) 2007 directive laying down minimum rules for the protection of chickens kept for meat production regulating chicken houses, stocking density, and surgical interventions.

(5) 2008 directive laying down minimum standards for the protection of calves regulating their conditions including limiting tethering and requiring that their accommodations allow each calf to lie down, rest, stand up, and groom itself without difficulty.

In light of the changes in technology, the more intensive confinement of animals used for food, and the public's increasing awareness and interest in the welfare of animals used for food, the Farm Animal Welfare Council

(FAWC), an independent advisory body established by the government of the United Kingdom in 1979, examined the effectiveness of the existing laws in the United Kingdom since the Brambell Report was published. The FAWC set out a strategy to improve the welfare of animals used for foods over the next 20 years. The FAWC report, *Farm Animal Welfare in Great Britain: Past Present and Future*, was published in October 2009. This report signaled an increased emphasis in the United Kingdom on improving the welfare of animals used for food. The following briefly summarizes their philosophy, findings, and proposed strategy for the next two decades.

The FAWC concluded that "at a minimum each farm animal should have a life that is worth living to the animal itself, and not just to its human keeper."[204] The FAWC admitted that the Five Freedoms have been the cornerstone of legislation to date and have been essential in improving the welfare of animals used for food in the past. However, the Five Freedoms focus on poor animal welfare and seek to relieve the suffering.[205] Instead, the law must now seek to ensure good welfare for these animals and move beyond merely protecting animals from cruelty and unnecessary suffering. The focus should be on setting a standard for the animal of a "life worth living," where the animal's wants (physiological, mental and behavioral) are addressed and the quality of the animal's life is, on balance, positive over the animal's lifetime, from birth to death.

Additionally, the government should aspire to provide the animals with a "good life" that goes well beyond a "life worth living," by establishing programs such as the Royal Society for the Prevention of Cruelty to Animals' (RSPCA's) Freedom Food scheme.[206] These programs require that farmers not only comply with the minimum requirements of the law, but also fully implement the codes of practice. The codes control animal disease by the strictest measures and with minimal prevalence; allow for normal behavior, availability of environmental choices and harmless wants; ban mutilations and certain husbandry practices; provide opportunities for an animal's comfort, pleasure, interest, and confidence; and meet the highest standards of veterinary care and stockmanship. The Five Freedoms, as amended to reflect a good life, would read that animals be provided food as a pleasurable experience and environments that animals seek out and enjoy, be given adequate anesthetic and pain relief for necessary surgery, be allowed no treatment that imposes a prolonged state of fear or distress, and be encouraged to engage in positive behaviors, such as play and social grooming.

The report recognizes that animal welfare is a function not only of legislation but also of husbandry practices and consumer demand.[207] While the law could be improved to impose higher standards and stricter enforcement, animal husbandry, veterinarian, and consumer efforts also are critical to achieving these goals. The FAWC includes several recommendations that focus on government support of welfare assurance schemes and regulation of consumer labeling, independent development of welfare measures, and assessment and reporting of welfare data.

The report states that eight conditions must be met for the new strategy to be effective in improving welfare.[208] These conditions include: (1) setting a legal standard that at a minimum every animal used for food has a life worth living and an expectation of enjoying a good life by incorporating the quality of life concept into the "Codes of Recommendations for Welfare of Livestock;" (2) improving surveillance and enforcement of the law; (3) educating stockmen on high standards of animal welfare and citizens about food and farming; and (4) maintaining a food supply chain with truthful labeling according to welfare provenance to provide consumer choice.

In addition, the government should strive to include in World Trade Organization treaties and negotiations that the welfare of animals used for food is a distinguishing characteristic of food, thereby allowing control over imports to offset market pressures. One major problem for farmers who implement more humane standards is being able to compete with those who do not. This is an issue for states within the United States, for countries within the EU, and for countries competing worldwide. The notion of free and unrestrained trade, whether it is within interstate commerce, inter-EU commerce, or international commerce, allows for some to take economic advantage of others. Providing animals used for food a good life will increase costs to the producer, and thus impact their sales. If a state has no ability to protect its own producers abiding by these heightened standards from imports that do not comply with them, its producers will suffer, as will their consumers. This is why consumer education and labeling is so critical, as are efforts to encourage a private trading system that rewards producers who enforce standards at or above the legal minimum.

The FAWC report is a good step forward in addressing the practical and legal criteria for improving animal welfare. However, "the devil is in the details." Many of the statements in the current law, as in the proposals for the future, sound good. But will they be effective? To some extent they suffer from the problems we have discussed throughout – the lack of specificity and use of terms that allow for unbridled balancing of costs and benefits relating to human and animal interests. The result is that in practice the human interests will continue to trump the animals' interests. Note that the broader the jurisdictional reach of the document, the more vague its terms will be. Thus, for example, the European Conventions that govern protection of animals under a variety of circumstances are designed to provide general guidance to the member states, and must be approved by all member states. In order to obtain approval of all the states and provide guidance rather than mandate express rules, the provisions include terms such as "within *reasonable* limits" and "*unnecessarily* put at risk." This allows for flexibility of interpretation and implementation, but also allows for members to do very little to protect animals. While states may draft more specific and higher standards, many states merely import the identical language into their laws and leave the courts or other enforcing entities the task of defining these terms. This makes enforcement quite difficult and provides very little protection for the animals in practice.

The directives are more explicit as they define an end result. They set out objectively measurable criteria and provide clear parameters for the member states to implement. The FAWC report highlights research into animal welfare quality and assessment undertaken by the EU Welfare Quality Project.[209] The goal of the project is to produce practical methods to assess the welfare of all animals used for food, including cattle, pigs, and chickens. Towards this end, the project has created a welfare assessment protocol that cites several welfare criteria, defines the welfare principles associated with the criteria, and sets forth examples of an assessment measurement.[210] Projects such as these are vital to providing true protection to animals used for food. It is important to note that the FAWC Council is advisory only, and often its advice is not heeded by the government. Nevertheless, the efforts in this area demonstrate some commitment to protect animal welfare on the farm. The United States could learn much about promoting the welfare of animals used for food from both the successes and failures of the United Kingdom and the European Union.

4
Animal Control and Management Laws

Domestic animals

In the United States, local jurisdictions enact laws pursuant to their police powers to protect the public health, safety, and welfare. Animal control laws, laws regulating the ownership of animals by citizens or unowned animals found within the local jurisdiction, are enacted pursuant to this authority and enforced by a local agency typically called "Animal Control." As is implied by the title, the goal of these laws is to "control" animals. In recent years, some jurisdictions have elected to modify the agency's title to "Animal Care and Control." Although the nature of the laws may not be significantly affected by the title change, indicating that the goal of the agency is not only to control animals but to provide for their care may alter the perception of the enforcers and the public towards animals as beings to be cared for and not merely controlled.

Controlling ownership

Animal control laws impose requirements and restrictions on the ownership of animals. For example, most jurisdictions require dog owners to obtain a license for their dog, to show proof of vaccination against rabies, and to not abandon or allow their dog to run at large.[1] Cat laws are less common, although some jurisdictions have similar laws for cats as well. Jurisdictions may also limit the number of animals a resident may own, and/or may require that the animals be sterilized to reduce the number of homeless and unwanted animals.

Enforcement of these laws can be difficult. For example, it is not uncommon for owners to fail to obtain a license for their dog and the authorities would not be aware of this unless the dog is left to run at large and seized by an officer. Since the dog is not licensed, animal control would not have a way to identify the owner. Thus, this dog is considered stray and will be held at the animal control shelter for a minimum number of days as prescribed under the law, generally around five days. If the owner claims the dog, animal control is then able to enforce the licensing provision and require as a condition to returning the dog that the owner obtain a license and otherwise comply with relevant laws. If the dog is not claimed by its owner, the animal becomes the property of the government and may be adopted to another owner or killed.

Challenging animal control laws

Citizens have challenged certain animal control laws as unconstitutional in violation of their right to due process. The following analyzes two such cases: citizens challenging running-at-large laws as applied to cats and "pet limit" laws. In both situations the citizens are arguing that the laws are unreasonable and/or arbitrary, interfering with their property interests in their companion animals, and failing to adequately serve governmental goals.

Running-at-large laws

While it is common for localities to require that dogs not run at large, it is less common for these laws to govern cats. Cats, even if feral, typically are not dangerous. Generally, they run away from humans and may attack only when cornered. Moreover, cats generally are not prone to leash training as are dogs, and thus requiring that they be on a leash when outside is not practical. Nevertheless, some jurisdictions prohibit cats from running at large and impound those who do. One such law from Akron, Ohio, requires that the cats be held for three days by animal control unless they are feral, ill, or severely injured, in which case they may be killed the day they are impounded. The reason for the law is to prevent a variety of "problems, including damage to property, cars, flower beds, shrubbery, gardens, and yards, damage caused by cats spraying on windows, health issues, and safety concerns related to cats darting into traffic."[2] Cat owners in Akron challenged the law as unconstitutional. They argued that many stray cats were harmed when trapped and others were deemed feral by animal control and killed immediately. As a result, their owners were unable to secure their return. Additionally, the use of trap-neuter-return (TNR) was a more effective means of addressing the alleged problems caused by unowned free-roaming cats, a population also targeted by this law. The plaintiffs were not successful.

First, the plaintiffs argued that the law was a violation of substantive due process as unreasonable and unduly oppressive to cat owners by interfering with their property rights without necessity. The court held that free-roaming cats caused the problems alleged and that trapping and killing them was a reasonable means to address the problem, even if some harmless cats are destroyed immediately. Moreover, even if TNR is more effective, the city need not use the most effective method available.[3] Second, the owners claimed their procedural due process rights were violated by killing their cats within 24 hours of impoundment without adequate notice. However, the court found that the plaintiffs failed to prove that healthy, non-feral cats had been killed before the three-day hold period was up and that providing an owner three days to reclaim his or her cat was sufficient. Owners could easily minimize the risk that their cats would be killed not only by keeping their cats indoors but also by microchipping them so that if found at large the owners would be notified if their cats were impounded.[4] Further, the plaintiffs argued that by giving traps to private individuals to trap free-roaming cats the government was violating the cat owners' Fourth Amendment right to reasonable search and seizure. The plaintiffs claimed that others would lure their cats off their property in order to trap them and bring them to the shelter to be killed. The court found this argument meritless as the facts did not support their allegation.[5] The plaintiffs failed to prove that citizens were luring cats off their owners' property to trap them.

Although the plaintiffs were not successful, this case highlights the problems with including cats under the running-at-large laws. First, the laws fail to account for unowned feral cats in a compassionate manner. Second, they fail to acknowledge the fundamental differences between cats and dogs both in terms of their natural instinct to roam and their more skittish nature around humans. This in turn creates a greater risk of being classified as "feral" when in fact the cat is only scared. Finally, citizens may trap cats but often not dogs, heightening the danger of improper seizure.

Pet limit laws

Some jurisdictions restrict the number of companion animals a person may own. Owners have alleged that these pet limit laws violate their right to due process because they are arbitrary and unreasonable. The following contrasts two state courts' views of such challenges.

In 1992, the Borough of Carnegie, Pennsylvania, enacted a law that stated in relevant part:

> WHEREAS, ... in the interest of preserving the health, safety and welfare of the residents of the Borough of Carnegie, Borough Council desires

to limit the number of dogs and cats kept by any one person and/or residence ... No person or residence shall be permitted to own, harbor or maintain more than five (5) dogs or cats, or any combination thereof, within the Borough limits ... Any person violating the provisions of this Ordinance shall be fined up to One Hundred ($100.00) Dollars plus costs for a first offense and up to Three Hundred Fifty ($350.50) [sic] Dollars for all subsequent offenses. Any violators of the provisions of this Ordinance are also subject to forfeiture of all animals over the five (5) pet limit.[6]

Soon after the law was enacted, Mary Creighton, who owned 25 cats, was cited for violation of the ordinance. Creighton appealed the violation arguing, inter alia, that the ordinance was arbitrary and unreasonable and thus violated her due process rights. Pursuant to the Fourteenth Amendment of the United States Constitution, the state may not seize an animal or otherwise restrict ownership of animals – for example property – without due process of law. Due process requires that laws be reasonably related to a legitimate government goal and provide adequate notice to citizens of what is required of them. Creighton argued that limiting companion animals to five was purely arbitrary and created a presumption that public health, safety, and welfare is based solely on the number of companion animals owned.[7] The court agreed, finding that while protecting public health is a legitimate government goal, the means sought was not necessarily reasonable, and remanded the case to the trial court. On remand the trial court was to determine if there was evidence to suggest that the mere number of animals in a residence is itself a threat to public health or that Creighton's residence was a nuisance or threat to public health.[8] Creighton also argued that enforcing this law against her violated the constitution as "an ex post facto law or an impairment of contracts and confiscates her pets without due process of law or just compensation,"[9] since she had the cats before the law was enacted. The court did not address this argument.

In contrast, a Minnesota court held that an ordinance limiting a household to two dogs was not unreasonable, and thus constitutional. That court stated: "it is the lack of a rational relationship between an ordinance and the general well-being, not the failure of the enacting city to consider empirical evidence that has caused ordinances to be struck down"[10] as unreasonable. Moreover, the owner challenging the law has the burden of proof and failed to demonstrate that limiting the number of dogs would *not* promote public well-being by reducing noise and odor.

Finally, while the number of pets allowed may be an arbitrary number, it does not render the ordinance itself arbitrary.

Arguably, whether constitutional or not, pet limit laws limit the potential for adoption of homeless animals and thus contribute to the death of homeless animals in our shelters. As the Pennsylvania decision implied, general laws regulating against nuisances and protecting public health already address the problem that may arise if an owner has too many companion animals and is unable to properly care for them. Some owners may not be able to adequately care for one animal while other owners may be able to properly care for dozens of animals. Instead of setting an arbitrary limit, the law should target the owners who create a public nuisance or neglect their animals.

Legislating to save homeless companion animals

There are hundreds of thousands of homeless companion animals in the United States. The fate of homeless animals in shelters is an extremely controversial issue. Many believe that animal shelters are to be truly shelters for homeless animals and that shelter animals should never be killed unless the animal is sick, injured, or truly dangerous and not able to be rehabilitated. Others argue that there are too many homeless animals, not enough homes, and the cost of caring for the healthy animals, much less rehabilitating the unhealthy ones, is prohibitive. Thus, the best animal control can do is to adopt out some animals and humanely kill the rest. It is estimated that 3.7 million animals were killed in U.S. animal shelters in 2008.[11] This number likely represents roughly 64 percent of all animals who entered shelters.[12] Laws that limit the number of companion animals one may own, or otherwise impose arbitrary restrictions on the ownership of animals, make adoption of homeless animals more difficult and contribute to this killing of innocent animals in our shelters.

Further, there are few, if any, laws that are designed to save the animals in our shelters from being killed unless the animal is "irremediably suffering or cannot be rehabilitated" or is truly "vicious" and a real threat to public safety. Laws designed to promote a "no-kill" agenda are extremely controversial. Most shelters claim they have the expertise and must be given the flexibility to deal with the animals brought to the shelter on a daily basis. Laws that impose restrictions on their decisions will result in chaos and animals languishing by the hundreds for years, a fate worse than death. They argue that the decisions that must be made hourly in an animal shelter cannot be "legislated."

No-kill advocates disagree. Advocates and lawyers have achieved no kill in several communities and the numbers are growing. Those who have been successful have drafted legislation that can achieve a no-kill community. The Companion Animal Protection Act (CAPA) provides a legislative framework for achieving a no-kill community.[13] Provisions of CAPA require that shelters:

1. provide animals with fresh food, fresh water, environmental enrichment, exercise, veterinary care, and cleanliness;
2. protect surrendered animals by
   (a) making them equally available for adoption or transfer as all other animals,
   (b) fully disclosing to all owners surrendering animals the likelihood their animal will be killed, and
   (c) refusing to kill any surrendered animal merely because the owner requests it;
3. allow volunteers to help with fostering, socializing, and assisting with adoptions;
4. not kill animals based on arbitrary criteria such as breed;
5. maintain contacts with and promote transfer of animals to all reputable rescues and no kill shelters or sanctuaries and not kill any animal when an alternative placement is available;[14]
6. have fully functioning adoption programs including offsite adoptions, use of the internet to promote their animals, and being open seven days per week for adoption;
7. spay/neuter all animals adopted and provide free spay/neuter for all feral cats and companion animals of qualifying low-income owners;
8. provide for trap-neuter-return of feral cats and protect feral cat caregivers; and
9. truthfully disclose to the public regularly the disposition of all animals impounded.

In May 2009, California Councilmember Portantino introduced Assembly Concurrent Resolution 74, requesting adoption of these no-kill principles by the state.[15] This resolution stands for the proposition that as sentient beings and cherished companions, homeless dogs, cats, rabbits, and other domestic companion animals, under the control of our government, should be legally protected from killing and euthanized only when in the best interests of the individual animal.

Dangerous dog laws

In addition to the general laws governing ownership of animals, most all jurisdictions have enacted "dangerous dog" laws designed specifically to protect the public from harm by dogs deemed "dangerous." The controversy surrounding these laws arises from their definition of "dangerous." Some jurisdictions target certain breeds of dogs that they believe are dangerous based purely on their breed and either ban them from their jurisdictions or impose severe restrictions on their ownership. Are these breed-discriminatory laws effective, efficient, fair, or constitutional? No.

Few empirical studies have tested the effectiveness or efficiency of laws that target dogs based on breed.[16] However, the studies that have been conducted demonstrate that breed discriminatory laws fail to keep the public any safer than nondiscriminatory laws and are very costly. For example, the United Kingdom banned pit bulls in 1991.[17] Researchers gathered data from one urban emergency department on dog attacks and compared the data over two periods: the three-month period prior to enactment of the ban and the three-month period two years after enactment of the ban.[18] The researchers concluded the ban had no affect on the number of dog attacks. More recently, a study conducted in Aragon, Spain, compared medically attended dog-bite statistics for the five-year period before the enactment of the city's breed ban in 1999 and the five-year period after the enactment. The researchers concluded that the "law was not effective in protecting people from dog bites in a significant manner."[19] Communities in the United States have had the same results from dog bans. A 2009 Denver newspaper article, written 20 years after enactment of Denver's pit bull ban, cited Denver's Animal Control Director admitting that he could not say with any certainty that the public was any safer due to the dog ban law, but could state that Denver citizens were less safe than Boulder citizens, where no dogs are banned and the dangerous dog law focuses on education and regulating dog owners.[20]

Studies have demonstrated that the breed bans are expensive as well. For example, Prince George's County, Maryland, established a task force to study the effectiveness of its pit bull ban.[21] The task force determined that the cost to the county over a two-year period was approximately $560,000, and concluded that the ban was "inefficient, costly, difficult to enforce, subjective, and questionable in its results."[22] The cost was a result of impounding hundreds of pit bulls a year, housing and caring for a large percentage of them during the lengthy hearing process, and eventually killing them because of their appearance. Meanwhile, hundreds of other dogs brought to the shelter were killed because there was no room for their care.[23] To better analyze the economic effect of breed bans, Best Friends Animal Society commissioned a study entitled "The Fiscal Impact of Breed Discriminatory Legislation in the United States."[24] For example, the model estimates that a breed-discriminatory law in the District of Columbia would cost $965,990 annually.[25] "The costs include those related to animal control and enforcement, kenneling and veterinary care, euthanasia and carcass disposal, litigation from residents appealing or contesting the law, and DNA testing."[26]

Generally, breed-discriminatory laws are upheld as constitutional when challenged in the United States.[27] The most successful challenges to breed-discriminatory laws focus on the law's definition of the breed and on the procedures for identifying and challenging the designation. Owners of pit bull-type dogs have argued that the definition of the breed targeted or banned as defined in the law is unconstitutionally vague and thus violates their right to due process. The challenged laws often define the breed as "American Pit Bull Terrier, American Staffordshire Terrier, and Staffordshire Terrier" and "any dog displaying a majority of the physical traits of any one or more of the stated breeds," and/or "any dog which has the appearance and characteristics of any other breed commonly known as a Pit Bull."[28] However, "pit bull" is not an official breed of dog and fundamental due process requires that the law be specific enough both to provide adequate notice to citizens of what is unlawful and to protect citizens from arbitrary enforcement of the law by government authorities. These definitions of the targeted breed fail to satisfy constitutional requirements because the determination of whether a dog has the appearance or displays the characteristics of a certain breed is highly subjective. The result is that the definition fails to provide adequate notice to the owner and results in arbitrary enforcement. Breed-discriminatory laws also will violate due process if adequate procedures are not provided to the owner. Such procedures include proper and timely notice that the dog has been deemed a member of the targeted breed and an opportunity to contest the breed classification of their dog in a proceeding that is fair and impartial.

Owners of banned breed dogs also have argued that breed-discriminatory laws are irrational and based purely on animus against the breed of dog. Several studies have shown that the breed of dog has less relevance to the dog's temperament or conduct than the treatment of the dog by the dog's owners or others. In fact pit bull-type dogs generally are highly obedient dogs and eager to please their human companions.[29] The commonalities involved in fatal dog attacks studied in 2006 by a researcher found that (1) 97 percent of the dogs had not been neutered by their owners; (2) 84 percent involved owners who either abused, neglected, chained, failed to constrain, or left their dog unsupervised around children; and (3) 78 percent of owners did not keep the dog as a companion but rather to guard, breed, or fight.[30] Thus, banning the breed will not rationally protect the public from harm.

Despite the evidence, courts are reluctant to find that breed bans are irrational. The only case to so hold was overruled on appeal.[31] However, in May 2009, the U.S. Court of Appeals for the Tenth Circuit held that a pit bull owner survived a motion to dismiss on a claim that Denver's pit bull ban is irrational and therefore unconstitutional.[32] The court stated that prior courts rejecting such claims had been decided decades before and thus could not have taken into account newly-acquired knowledge concerning dog breed and temperament. Moreover, the American Kennel Club and United Kennel Club standards support the plaintiff's claims that pit bulls are not dangerous per se. These standards state: "American Pit Bull Terriers make excellent family companions and have always been noted for their love of children," and that the "Staffordshire Bull Terrier ... with its affection for its friends, and children in particular, its off-duty quietness and trustworthy stability, [is] a foremost all-purpose dog."[33] Thus, the plaintiff was free to proceed to develop the record towards a final decision of whether the law is, in fact, irrational.

Many other jurisdictions follow a different path to protecting the public from "dangerous dogs" by enacting laws that regulate conduct rather than target dogs based on breed. But whose conduct should the law regulate – the dog's or the owner's? One typical approach to dangerous dog laws is to define dog behaviors that would classify the dog as either potentially dangerous or dangerous and then impose restrictions on their ownership and handling. For example, the United Kingdom dangerous dog law makes it an offense for a dog to be "dangerously out of control in a public place."[34] A handbook explains that a dog is "dangerously out of control if it injures a person, or it behaves in a way that makes a person worried that it might injure them."[35] This is a difficult standard to apply as it requires an owner to anticipate what another person feels in response to their dog.

In comparison, Washington DC's law is typical of laws that define dangerous dogs and their conduct more explicitly. The DC law defines a potentially dangerous dog as:

> any dog that: (i) Without provocation, chases or menaces a person or domestic animal in an aggressive manner, causing an injury to a person or domestic animal that is less severe than a serious injury;[36] (ii) In a menacing manner, approaches without provocation any person or domestic animal as if to attack, or has demonstrated a propensity

to attack without provocation or otherwise to endanger the safety of human beings or domestic animals; or (iii) Is running at-large and has been impounded by an animal control agency 3 or more times in the District within any 12-month period.[37]

A dangerous dog is defined as:

any dog that without provocation: (i) Causes a serious injury to a person or domestic animal; or (ii) Without provocation, chases or menaces a person or domestic animal in an aggressive manner, causing an injury to a person or domestic animal that is less severe than a serious injury [subsequent to having been determined to be a potentially dangerous dog].[38]

Courts have addressed the interpretation of "provocation" in these dangerous dog laws. In Illinois, for example, courts have held that provocation should be determined from the perspective of the reasonable dog.[39] In other words, the question is whether a normal dog would have responded to the events that led up to the confrontation in a similar manner as the defendant dog. Thus, consider the case where a young child accidentally stepped or fell on the tail of a dog chewing a bone and the dog scratched the child's eye. The court held that the dog was provoked and the dog's reaction was that of a reasonable dog.[40]

Once the dog is classified as potentially dangerous or dangerous under these laws, a variety of restrictions are imposed. In DC the owner must specially register the dog, which requires that the owner be 18 years or older, have written permission of the property owner or homeowner's association where the dog resides, and pay a fee. In addition, the dog must be licensed, have current vaccinations, be spayed or neutered and microchipped, and be confined in a proper enclosure.[41] The only difference in restrictions between a potentially dangerous dog and a dangerous dog is that a potentially dangerous dog may leave the proper enclosure "when under the control of a responsible person and restrained by a chain or leash, not exceeding 4 feet in length," whereas a dangerous dog must remain confined in the enclosure and on the owner's property except when needed for medical attention, and then the dangerous dog "shall be caged or under the control of a responsible person and muzzled and restrained with a chain or leash, not exceeding 4 feet in length."[42] Owners who violate these provisions are guilty of a misdemeanor and subject to a fine not to exceed $500 for a first offense or $1,000 for subsequent offences and/or imprisonment not to exceed 90 days.[43] The dog may be humanely destroyed if it is deemed to be a threat to the public if returned to the owner, or if the owner either violates any of the aforementioned provisions, fails to reimburse animal control for costs of the dog's confinement, or forfeits the dog for destruction.

While these laws are not breed-discriminatory but rather based on the conduct of the dog, they target the dog even though the dog may not be the truly liable party. The sanctionable conduct is that of the dog and the primary sanctions are imposed upon the dog. Moreover, the sanctions are quite severe from the dog's perspective: a life imprisoned in an enclosure and perhaps death. One problem with this approach is that the dog does not know what the law states and thus has no notice of what is expected of him or her. Not only is this unfair, it is ineffective to deter the dog's wrongful conduct. Instead, the law should target the individual with notice of the law and often the truly responsible party – the dog's owner or handler.

As indicated above, the factors common to the large majority of dog-bite cases involve owners who abuse, neglect, chain, improperly supervise, or fail to neuter their dogs, as well as those who keep their dogs for purposes other than companionship.[44] Some jurisdictions are addressing these factors directly with laws that target these reckless owners. For example, "reckless owner" laws include the following provisions – convicted felons may own only sterilized dogs, owners cited more than once for animal abuse or neglect are prohibited from owning a companion animal, and owners who repeatedly violate animal control laws are deemed "problem pet owners" and may be forced to surrender their animals. Jurisdictions restrict chaining or tethering dogs when unattended, and most have leash laws. Other jurisdictions focus their efforts on public education and stiff fines on the owner. For example, if a dog bites a person, a $350 fine is imposed on the owner, and if the injured person requires medical attention the fine is $750. If a dog attacks another dog or is not licensed, the fine is $250.

Just as criminal laws protect citizens from physical harm at the hands of other humans and punish those who violate the law, animal control laws protect citizens from harm by dogs. The approach chosen is critical both to the effectiveness and the fairness of the law. While it is true that humans have bred dogs to have certain physical and behavioral traits, the evidence suggests that breed is not determinative of a dog's temperament. Thus, discriminating against a dog merely because of its breed is not unlike discriminating against a human because of his or her race, if breed, like race, is irrelevant to whether the dog is dangerous.

Breed-discriminatory laws are also under-inclusive and over-inclusive. Reckless owners own a variety of breeds of dogs and their dogs could be dangerous independent of their breed. Moreover, not all dogs of the same breed have the same behavioral traits and dogs of the same breed are owned by all sorts of owners – some reckless, but many responsible and loving. Thus, not all dogs of the same breed are dangerous. By regulating the owner's conduct rather than targeting the dog, the law can more effectively protect both the public from harm and the dog from improper care and handling at the hands of the owner. A nondiscriminatory approach focusing on reckless owners is as important to protect the interests of animals as it is to protect their owners. Animals, as persons, are individuals. A law that targets conduct and not breed (or race or sex) recognizes that all sentient beings are individuals and regulates them based on their conduct, not on their breed, race or sex. Moreover, the responsible party's conduct must be targeted. Thus, when the animal's owner is the blameworthy party, the owner should be held responsible and punished, not the innocent animal.

A hypothetical case study: feral cats

Caught between two worlds

> Just like feral cats occupy a unique niche between wild and domestic, they also occupy a gray zone in the law. For many cats, their status as "domestic" animals means certain death in shelters. But wild animals tend to fare little better.
>
> In those states where it is allowed, wildlife is subjected to trapping, poisoning and hunting, particularly if they are an unprotected species. Feral cats, in essence, are caught between two anachronistic world views. If they are legally domestic, they are subject to mass slaughter in shelters by the humane movement. If they are legally wild, they are subject to killing by hunting, trapping, and poisoning.[45]

Feral cats are cats who have escaped from their home, live on the street, and have become unsocialized to humans, or were born on the street and were never socialized to humans. As a result, these cats are "wild." Stray cats are cats found at large but are socialized to humans. Often it is difficult to distinguish a stray cat from a feral cat because when captured they may act very much alike and appear "wild" to the person who captures the cat. The law often will include feral cats in the definition of stray cat. In a few jurisdictions, stray and/or feral cats are deemed "wildlife" and are regulated under the state game laws. For example, in Wyoming, stray cats are listed with coyote, raccoons, skunks, and porcupines as "predatory animals," are excluded from protection, and may be "taken without a license in any manner and at any time."[46] The Commonwealth of New South Wales, Australia, similarly passed a Game Act which defines cats "living in the wild" as a "game animal" considered a "pest" and may be hunted on private lands without a license.

In most jurisdictions, although feral cats act "wild" they are not considered wildlife because they are a domesticated species. Thus, they fall under the domestic animal control laws that are designed to govern animals socialized to humans and often do not have provisions for feral cats. This creates problems for the feral cats, their caretakers, and society. There are no published decisions found addressing these issues. Therefore, the following is a hypothetical story of a woman, Joanne, and a feral cat, Liberty, and how the law will treat their situation using Prince George's County, Maryland, as a sample jurisdiction.

Joanne is a middle-aged woman who likes cats. She began to see cats who lived on her street but who did not appear to "belong" to any human. When she would approach them they would run off but they would return when she went into her house. They appeared hungry so Joanne started leaving food out for them in the mornings and evenings. The food quickly disappeared within a couple of hours, although she rarely saw the cats eating. After a few months, Joanne noticed four small kittens with one of the older cats. One of the adult cats, whom she named Liberty, had had a litter of kittens and now the whole family came by to eat. Joanne realized that the situation would get out of control when the adult cats, and soon their kittens, would have more kittens.

In fact, the stray/feral cat population is a serious problem in many urban areas. Hungry cats abandoned by their owners are attracted to a food supply. In urban areas, trash is often left in the streets and alleys. The hungry cats create a home or colony around these areas where they eat, roam, and mate. The traditional solution under the law has been for animal control to trap the cats, bring them to the shelter, and kill them. However, studies have shown that this does not work to eradicate the stray/feral cat population because of a process known as "compensatory reproduction." As the number of feral cats in a colony or area is reduced, the remaining cats breed at younger ages and have less time between litters. Thus, feral/stray cats compensate for the reduction in members of a colony by increasing their breeding practices, making up for their lost companions who were trapped and killed.

Animal advocates argue that a more compassionate and effective solution is a procedure known as trap-neuter-return (TNR). Interested citizens go to the location of a colony and trap the cats in the colony. They bring the cats to a veterinarian for neuter and vaccination. The vet also ear-tips the cats – removing the tip of the left ear of the cat while under anesthesia. This is the universal sign that the cat is feral and has been neutered and vaccinated (at least once). Under ideal circumstances the cats would also be microchipped. After 24–48 hours the cats are returned to the same location to live out their lives without reproducing. Under ideal circumstances, a neighbor assumes the role of caretaker for the colony by creating an outdoor shelter, providing food and water for the cats, and monitoring the colony for sick or injured members or new members who must then be TNR'd to prevent reproduction.

Is TNR legal?[47] If Joanne decided to trap, neuter, and return Liberty and her brood to the street and continue to care for them, would she be subject to liability under the law? In this sample jurisdiction of Prince George's County, the answer is yes, IF Joanne is deemed an "owner" of the cats.

According to the law of Prince George's County, Liberty and the others in the colony are "animals" defined as "every nonhuman species of animal, both domesticated and wild, including, but not limited to, dogs, cats, ferrets, livestock, and fowl."[48] But what type of animal is Liberty? The law defines the following classes of animals:

Cat shall mean domesticated felines. The term "cat" shall not include wild or exotic felines.

Wild animal means any animal which is not included in the definition of "domesticated animal" and shall include any hybrid animal which is part wild animal.

Feral shall mean animals existing in the wild or untamed state, i.e., wild.

Domesticated animal means an animal of a species that has been bred, raised, and is accustomed to live in or about the habitation of man, and is dependent on man for food or shelter.

Companion animal shall mean any domestic or feral dog, *domestic or feral cat,* ferret, nonhuman primate, guinea pig, hamster, rabbit not raised for human food or fiber, exotic or native animal, reptile, exotic or native bird, or any feral animal or any animal under the care, custody, or ownership of a person or any animal which is bought, sold, traded, or bartered by any person.[49]

Arguably, Liberty is a domesticated, feral, companion cat. She is not a "wild animal," even though she is "feral," because "wild animal" does not include "domesticated animal," and the definition of a domesticated animal is based on the nature of the species, not on the nature of the individual animal.

Does Joanne "own" Liberty? Under the law, an owner is any person who "has a right of property in an animal; keeps or harbors an animal; has an animal in his or her care; [exercises control over an animal; or] acts as a temporary or permanent custodian of an animal."[50] The law further defines "keeping and harboring" as "the act of, or the permitting or sufferance by, an owner or occupant of real property either of feeding or sheltering any domesticated animal on the premises of the occupant or owner thereof."[51] Since Liberty is considered a domesticated animal, Joanne would arguably fall under the definition of "owner" since she arguably "keeps and harbors" Liberty by providing her with food on her premises.

As a result, both Joanne and Liberty are in trouble under the law. Here are the consequences for Joanne. First, Joanne must obtain a license for Liberty and all of the other cats in the colony – a difficult and somewhat costly requirement. Second, Liberty is considered a "stray," which is defined as "any animal found roaming, running, or self-hunting off the property of its owner or custodian and not under its owner's or custodian's immediate control"[52] and an "animal at-large" defined as "an animal not under restraint and off the premises of its owner."[53] The law makes it unlawful for an owner to permit an animal to be "at large." Thus, Joanne is in violation of the at-large law. Third, any animal found at large or running at large is declared to be a nuisance and "dangerous to the public health, safety, and welfare."[54] Joanne is thus in violation of a second provision that prohibits nuisances. Fourth, Joanne is also liable for failing to remove Liberty's excrement. Joanne would not be liable for abandoning Liberty because she remains as a caretaker for her and the rest of the colony. However, in some situations, citizen trappers go to neighborhoods to perform TNR but do not live close enough to be caretakers of the colony. If there is no person designated as their caretaker, the trapper would be liable for abandoning the cat.

As for Liberty, since she is stray, found at large, and considered a nuisance, animal control is authorized to impound her.[55] Unless Joanne is notified that Liberty has been impounded, a highly unlikely proposition, Liberty becomes the property of the jurisdiction and subject to adoption or killing. However, since Liberty is not socialized to humans, she is by definition not "adoptable" and thus will be killed likely within 24 hours.

The bottom line is that Joanne has three options: she can choose to feed Liberty and the other cats and risk being held liable for violating several provisions of the law; she can ignore the cats; or she can call animal control to impound and kill them. None of these choices are good for Joanne or Liberty.

The compassionate legal solution

Compassionate communities enact laws to provide a bridge for Liberty and her feral companions between the two worlds of wild and domestic animals and to allow Joanne to care for them without being subject to liability. These laws specifically recognize TNR as a program to address the feral cat population without trapping and killing the cats. TNR laws may include the following components.[56] First, defining "feral cat" as a cat who is *"free-roaming*, unsocialized to humans and *unowned."* Second, defining "feral cat caregiver" as "someone who cares for cats and has an interest in protecting cats but is *not the owner* of those cats." Third, granting caregivers "the same rights of redemption for a feral cat as an owner of a pet cat without conferring ownership of the cats on the caregiver." Fourth, requiring that animal control "promote trap, spay/neuter, and return (TNR) as the means of reducing and eventually eradicating feral cats and protecting the public." Fifth, prohibiting animal control from "providing traps to the public to capture cats except to a person for the purpose of catching and reclaiming that person's cat(s), to capture sick or injured cat(s) or cats otherwise in danger not including the mere location of the cat, to capture feral kittens for purposes of taming and adoption, or, for capturing feral cats for purposes of TNR." Sixth, subjecting a "person who uses a trap to trap any cat that resides in the colony of a caregiver for purposes other than those listed" to a fine. In addition to these basic provisions, some communities require that a caregiver be available and registered with animal control in order for the TNR to be lawful. Other communities allow TNR even if there is no registered caregiver but do not provide free veterinary care for the neutering and vaccinations unless the caregiver is registered.

Enacting a law that specifically addresses feral cats in a compassionate manner and allows them to live out their lives naturally demonstrates a recognition and respect for feral cats as sentient beings with an interest in life. Moreover, it provides protection for the cats' needs while reducing their numbers without mass killing. Finally, the costs associated with eradication programs far exceed that of TNR programs. For example, the Best Friends TNR Cost Savings Calculator, based on a model developed by John Dunham & Associates, estimates that 102,340 feral cats reside in the District of Columbia and the city would save approximately

$4 million addressing the feral cat problem by using TNR rather than trap-and-kill methods.[57]

Cats versus native species prey

TNR laws are very controversial and not only with people who do not like cats or find them to be a nuisance because they urinate on their flowerbeds. Bird enthusiasts and environmentalists often are opponents of feral cats. There is a long-standing feud between animal advocates – those who believe that feral cats have a right to live out their lives and favor TNR – and those who believe that feral cats may eradicate certain species of birds or other small wildlife if allowed to live and roam freely. A 2009 article on TNR and its effects on native species published in *Conservation Biology* states: "Domestic cats are on the list of the 100 worst invasive species globally."[58] The authors caution: "conservation scientists and advocates must properly identify the environmental implications of feral cat management ... on species conservation, the physical environment, and human health." According to the authors, TNR can be devastating to the environment because feral cats, subsidized by humans, are generally found at 10–100 times higher densities than similarly sized native predators. Cats, whether tame or feral, instinctually hunt small prey. As a result, the higher density of even well-fed feral cats will affect the population of small native species, and thus impact on the area's environmental balance.[59] On the positive side, however, many communities favor feral cats as a pest control device to kill rats.

TNR advocates present a very different picture of the environmental impact of feral cats in bird species populations. TNR advocates note that feral cats, given their long presence in ecosystems, do not contribute significantly to the decline of native species. In fact, those who claim that cat predation is among the top causes of bird species loss do so by "characterizing 'habitat destruction' as a single cause of species loss [that in turn] belies the vast human impact encompassed by the term. Logging, crop farming, livestock grazing, mining, industrial and residential development, urban sprawl, road building, dam building, and pesticide use are just a few of the hundreds or even thousands of activities and damages that are captured by this phrase."[60] Once these human activities are lumped together, other relatively minimal causes of species loss, such as cat predation, appear incorrectly to be significant threats.

Moreover, "[k]illing a population of any adaptive species [like feral cats], while leaving in place the advantageous habitat in which it thrives, merely leaves an ecological void which is quickly refilled by members of that species."[61] This is demonstrated by the fact that although millions of stray and feral cats are killed every year in U.S. shelters, the stray and feral cat populations thrive.

These advocates conclude that human activities are the primary cause of the loss of bird species. "Using a simplistic and fallacious cat versus bird argument to set policy comes at the cost of millions of animal lives – not only of the cats who shoulder the blame of our human mistakes, but of the very birds these individuals aim to protect."[62]

Wildlife

State game laws

Many states have an entity empowered to establish laws that regulate wildlife within the state. For example, Florida, with a 1999 constitutional amendment, created the Florida Fish and Wildlife Conservation Commission. Its stated mission is "[m]anaging fish and wildlife resources for their long-term well-being and the benefit of people."[63] The commission enacts rules and regulations that govern captive animals, boating, hunting, fishing, and trapping.

Private ownership of captive wildlife

The captive animal regulations define categories of wild animals who may be personally owned and the types of permits needed.[64] For example, Class I wildlife[65] are those animals who pose a significant danger to people, and personal possession is illegal unless they were possessed on or before August 1, 1980, when the law was enacted. Class II wildlife[66] can also pose a danger to people. Special permits are required for public exhibition, sale, or personal possession. All other wildlife are Class III wildlife and require a standard permit. Possession of Class I and II wildlife require substantial experience and specific cage requirements.

Allowing private ownership of wildlife is extremely controversial, not only from a public safety standpoint but also from the standpoint of the animal. For example, in December 2009, a 40-pound serval was reported missing by its owner in St. Paul, Minnesota, and ultimately captured by animal control. In August 2009, an elderly couple was found dead from unknown causes in a house filled with dogs, cats, birds, and exotic monkeys. In August 2003, a 400-pound Bengal tiger and a three-foot alligator were seized from a New York apartment and the owner charged with reckless endangerment.[67] Many believe it is unethical to keep a wild animal, such as a monkey, cougar, cheetah, alligator, or Bengal tiger, caged in a home. Clearly the animal is not in its natural habitat, would have little, if any opportunity to engage in natural activities, and has no companionship with its own species. Keeping an animal under these conditions disrespects its inherent interests in order to satisfy the personal gratification of an individual human. For these, and public safety reasons, several jurisdictions forbid ownership of most wild or exotic animals.[68]

Hunting, fishing, and trapping

The hunting and fishing laws define when citizens may hunt or fish, often referred to as "open season."[69] A license is generally required, for a fee, although often there are no requirements of training or other education. Such laws also may define the methods of hunting and trapping that are permitted under the laws, including the use of guns, bows and arrows, leg-hold traps, and snares, as discussed in Chapter 2.

Sport or recreational hunting is decreasing throughout the United States. Hunting is defined as the pursuing or chasing of game or other wild animal. The National Shooting Sports Foundation (NSSF) website states: "Surveys show nearly 80 percent of Americans support hunting, although less than 10 percent actually participate. These 18.5 million hunters contribute more than $30 billion annually to the U.S. economy and support more than 986,000 jobs."[70] Those who support hunting claim it is an inexpensive method to manage wildlife, provides economic benefits to rural towns and sporting goods stores, and is a cultural tradition. The use of sport hunting to manage deer populations is quite prevalent. For example, in Florida the commission aligns open season on white-tailed deer with peak rutting (breeding) activity, stating that deer hunters want to hunt during the rut as "bucks are more active during daylight (shooting) hours and often less cautious, making the chances of seeing them in the field better."[71]

Hunting is also highly controversial. Opponents argue that sport hunting does not reduce deer populations because the remaining animals flourish due to less competition for food, decreases in density, and other factors. The result is that compensatory reproduction occurs; namely reproduction rates increase, adding to the alleged overpopulation as a result of the hunt. The fact that the majority of deer killed are male also encourages an increased herd size since it skews the normal male-female ratio so that an abnormally large proportion of deer herds in heavily hunted areas are reproducing females. Opponents also argue that sport hunting is unethical because of the cruelty inflicted upon the animals who either are injured but not killed or who suffer in pain until they ultimately die. In addition, the killing of sentient beings for pure sport or recreation is unjustifiable. Finally, sport hunting disturbs the social structure of the deer herd, often causing abnormal behavior which in turn may create additional environmental damage. According to many social scientists and psychologists, hunting may lead to a lack of empathy and sympathy, especially in young children, and thus to anti-social behavior.[72] Furthermore, increased rates of homicide and suicide may be due to the increased availability of guns.[73]

The controversy over sport hunting and animal activists' attempts to promote their message by interfering with hunters during a hunt triggered the enactment of state hunter harassment laws. Hunter harassment statutes were enacted by a majority of states in the United States between the years 1981 and 1996 through efforts of the Sportsmen's Caucus and the Wildlife Legislative Fund of America, a pro-hunting lobby based in Washington, DC.[74] These statutes have been challenged on the ground that they violate animal advocates' right to free speech as they are merely demonstrating their contempt for the immoral sport of hunting. Interestingly, although the United States First Amendment protection of speech is likely the strongest in the world, the United States enacted these laws while such laws are unheard of in many other countries, including the United Kingdom, Spain, and other countries of the European Union.

A few of the early statutes were found unconstitutional as either content-based or overly broad and vague. However, they were easily amended to bring within constitutional limits, according the courts. For example, the Minnesota Court of Appeals in *State* v. *Miner*[75] construed the Minnesota hunter harassment statute to be impermissibly content-based. The statute then stated in part:

> A person who has the intent to prevent, disrupt, or *dissuade* the taking of a wild animal or enjoyment of the out-of-doors may not disturb or interfere with another person who is lawfully taking a wild animal or preparing to take a wild animal.[76]

"Dissuade," the court explained, "carries a connotation of using argument, reasoning, entreaty, admonition, advice or appeal to convey a message. The statute therefore discriminated between opposing points of view by attempting to silence persons intending to convey a particular message."[77] As such, the law cannot stand as it is not necessary to promote a compelling government interest. The Supreme Court of Illinois found the same problem with its statute, excised the provision, and found the remainder of the statute constitutional.[78]

The Second Circuit in *Dorman* v. *Satti*[79] struck down the first Connecticut hunter harassment statute as unreasonably overbroad and vague as it prohibited persons from interfering with another person who was taking or *acting in preparation* to take wildlife. The court stated that an "act in preparation" is nowhere defined in the statute and thus the act reaches a wide range of activities confined to no particular time, place, or manner. The "acts in preparation" clause can be reasonably read to encompass "buying supplies long before the actual hunt takes place ...; consulting a road map ...; making plans during a workplace coffee break; or even getting a good night's sleep before embarking on a hunting trip."[80] Accordingly, the Connecticut Hunter Harassment Act was unconstitutional on its face.

Soon after this case, Connecticut amended the law. The current Connecticut Hunter Harassment Act reads:

(a) No person shall obstruct or interfere with the lawful taking of wildlife by another person at the location where the activity is taking place with intent to prevent such taking.
(b) A person violates this section when he intentionally or knowingly:
(1) Drives or disturbs wildlife for the purpose of disrupting the lawful taking of wildlife where another person is engaged in the process of lawfully taking wildlife;
(2) blocks, impedes or otherwise harasses another person who is engaged in the process of lawfully taking wildlife;
(3) uses natural or artificial visual, aural, olfactory or physical stimuli to affect wildlife behavior in order to hinder or prevent the lawful taking of wildlife;
(4) erects barriers with the intent to deny ingress or egress to areas where the lawful taking of wildlife may occur;
(5) interjects himself into the line of fire;
(6) affects the condition or placement of personal or public property intended for use in the lawful taking of wildlife in order to impair its usefulness or prevent its use; or
(7) enters or remains upon private lands without the permission of the owner or his agent, with intent to violate this section.[81]

The Supreme Court of Connecticut in *State v. Ball*[82] upheld this statute as constitutional. The constitutional test proceeded as follows.

Is the regulation of speech content-neutral or content-based?

The Connecticut Supreme Court held the statute content-neutral because the government is not targeting the content of the message itself. The court explained that the law merely prohibits disturbing the hunter engaged in a lawful hunt. If he or she disturbs the hunter, it is a violation, independent of what the protestor says. Further, the statute was not enacted because of disagreement with the message conveyed but rather to maintain safe hunts and control the overpopulation of deer, which itself is unrelated to the content of the protestor's message.

In contrast, Judge Harrison of Indiana, dissenting in a case challenging a similar Indiana hunter harassment statute, found the statute was content-based because culpability was limited to those whose intent is to prevent the taking of wild animals.[83] A statute is deemed content-based when the government adopts a regulation of speech because of disagreement with the message it conveys. He stated: "[t]he law affects one group and one group only: those who are morally or philosophically opposed to capturing or killing animals."[84] Moreover, the sanctions were severe and included punitive damages to the hunters whose efforts they thwarted. The only purpose of such damages is to "protect the personal, pecuniary interests of sportsmen and quell dissent by those who oppose the hunt."

Is the act properly tailored to serve an appropriate government goal?

If content-neutral, the act must be "narrowly tailored to serve a significant government interest, and leave open ample alternative channels of communication."[85] The Connecticut Supreme Court first found that the state's identified interests in public safety, including reducing confrontations between hunters and activists and deer-vehicle collisions, raising revenue, managing wildlife, providing recreational opportunities for its citizens and protecting their right to hunt are significant. The court explained that:

> The state offered particularly compelling evidence relative to the role of hunting as a forest and wildlife management tool. Increased development in the state ... has resulted in an ecosystem that is no longer self-balancing. In particular, the deer population is not naturally controlled by predators, leaving regulated hunting as the most cost efficient, effective means of controlling the deer population, according to the collective experience of the fifty states. The control of the deer population, in turn, has a trickle down effect on many other areas of significant state interest. It prevents overbrowsing in state forests and on state residents' ornamental plantings. Indeed, a state biologist testified that an uncontrolled deer population could strip forests of their ground growth and young trees in a matter of decades. Hunting also helps to control the population of geese, coyotes and a variety of small game.[86]

Second, the act is narrowly tailored because it is limited to prohibiting interference with the lawful taking of wildlife "at the location where the activity is taking place with intent to prevent such taking."[87] Moreover, although the act broadly addresses the many ways in which that interference might occur, each section of the act is strictly limited to actions that are intended to interfere with, or actually interfere with, the lawful taking of wildlife. Finally, there are other alternative means of communication left open at other times and locations for the protestors to present their message.[88]

If a statute is content-based, the act must be necessary to serve a compelling state interest and narrowly drawn to achieve that end. Judge Harrison, after finding the Illinois statute content-based, held the Illinois statute failed this test. He explained that:

> sportsmen may be entitled to engage in lawful hunting, but they do not have the right to do so free from annoyance, harassment and confrontation. Hunting and the treatment of animals present important moral, social and political issues. For some, those issues are every bit as compelling as racial equality, gender discrimination, and abortion. Under the first amendment, we cannot circumscribe the debate on those questions any more than we can limit the debate on the rights of women and minorities or the state's role in regulating reproduction. As in all of those areas, opposition to the protestors' views may be strident, but "the fact that society may find speech offensive is not a sufficient reason for suppressing it. Indeed, if it is the speaker's opinion that gives offense, that consequence is a reason for according it constitutional protection."[89]

Moreover, there is no reason to suspect that the hunter will kill another human even if he or she is willing to kill an innocent animal. But even if this is a concern of the state, the state must protect the protestor, not silence him or her.

Similarly, hunting in Australasia is popular and is conducted for a number of purposes – for recreation alone, for sport, trophies, sustenance, "pest" control, and commercial gain. The legal framework in Australasia is similar to that in the United States. A scholar, after reviewing the laws of recreational hunting in Australasia, stated that "none could be said to provide a robust protection regime."[90] She summarized the problems with the legislation as failing to require any justification for hunting, vaguely worded provisions such as "usual and reasonable manner and without causing excess suffering," and relying "on extrinsic documents such as codes of practice which ... are largely unenforceable." [91]

Urban wildlife control

Wildlife in urban settings pose unique problems for local governments. These animals are generally referenced as "nuisance wildlife" or "pests," and management of them is often referred to as "damage control." The terms themselves demonstrate what policy-makers think of these sentient creatures; clearly, they have no interest in protecting them. Wildlife damage control operators make a profit from killing these animals and are now commonplace in large U.S. urban settings with little government oversight. The numbers of wild animals trapped and killed by these operators, including opossums, squirrels, raccoons, bats, and other species, typically run into the tens of thousands in states where they have tracked the numbers. The methods used to kill the animals are often inhumane. Cage traps, for example, often allowed, frequently expose the animal to extreme heat or cold if the animal is left in the trap for several hours. Animals that seek shelter in attics or crawl spaces often do so to give birth, and removing the adult animal leaves their young to die. Furthermore, if there are places to relocate the animals, relocation may lead to increased risk of predation from other species.

Several states regulate wildlife control operators.[92] Maryland, for example, regulates wildlife damage operators in order "to protect public health and safety while conserving diverse wildlife populations."[93] Maryland requires a person to obtain a permit before engaging in any commercial[94] wildlife damage control. Such a permit requires passing with a grade of 80 percent on a written examination that tests the individual's knowledge of wildlife biology, legal lethal and non-lethal control methods, acceptable methods of killing, human health and safety issues, and other aspects of wildlife damage control. The law prohibits the killing of limited species and otherwise allows for the animal to be released on site, transported to a wildlife management area, or killed under proper methods. As of early 2010 the District of Columbia was considering a bill to regulate wildlife control operators that will provide additional protections to wildlife, including (1) allowing control "only if an animal ... is causing actual damage to property or posing an immediate health or safety threat to persons or domestic animals;" (2) allowing "lethal control ... only when public safety is immediately threatened or when nonlethal control methods have been employed to address the specific problem at the site and have proven unsuccessful;" and (3) prohibiting the use of glue traps, leg-hold and other body-gripping traps, body-crushing traps, snares, or harpoon-type traps.[95] It is unclear how damage control operators will be able to determine when an animal is causing an immediate health threat or when public safety is immediately threatened.

Another related urban wildlife management issue concerns white-tailed deer.[96] Deer are found throughout the United States, including in urban cities. State game commissions have responsibility for managing wildlife populations, including deer. Population management of urban and suburban populations of deer presents special problems for the government. Homeowners complain that the deer are eating their flowers and shrubs, and that deer-vehicle collisions are a serious safety threat to humans and deer. The traditional and most common method of deer management is to hunt, kill, and remove the deer. There is significant controversy over whether this method is humane, effective, and efficient. In an urban setting, however, hunting is particularly problematic. Sport hunting often is not an option because of safety concerns, and thus sharp-shooters or bow-hunting alternatives may be utilized. However, these alternatives may be unwise, unsafe, and/or publicly unacceptable as inhumane.

While many believe that killing the deer remains the best method of population management, others claim that it is counterproductive due to compensatory reproduction. They argue that the better solution is to address the problem – reproduction, and not the symptom – overpopulation. An immunocontraceptive, Porcine Zona Pellucida, commonly known as PZP, has been effectively utilized to manage deer in urban settings. Studies have demonstrated that PZP programs can reduce deer populations by up to 10 percent per year. One such project was conducted for the U.S. Department of Commerce's National Institute of Standards and Technology, in Gaithersburg, Maryland.[97] The project began in 1996 with an initial inoculation in the spring.[98] During the following four years, 1997–2001, the number of births was cut by approximately 72 percent. Concurrently, the deer population was reduced and stabilized. There were nearly 300 deer in 1997. In January 2009, the number hovered around 200 with an expectation of a significant reduction by 2014 as all of the does were to be vaccinated and many of the resident deer were to reach their natural life-span. The Gaithersburg data also indicated a decrease in deer-car collisions coincident with the reduction in deer population as a result of contraception.[99] Data from Fire Island National Seashore also demonstrate the efficiency of PZP in reducing deer numbers. "Between 1993 and 1997, fawning rates among individually known, treated adult females decreased by 78.9% from pre-treatment rates." After an initial increase, population density in the most heavily treated areas "decreased at 23% per year [from 1998] to October 2000.[100] In a study on Fripp Island, South Carolina, preliminary data on deer showed the "efficacious transition [for PZP] from a two-injection contraceptive lasting 1 year to a one-injection contraceptive lasting 2 years."[101]

GonaCon is another contraceptive that has been developed by government scientists with the United States Department of Agriculture. The slaughter of deer as a means of population control is ineffective and contrary to humane treatment. Animal protection lawyers should work on addressing the problems encountered when human populations intrude on the habitat of wildlife or when state game associations deliberately foster the increase of so-called game animals at the expense of other animals and often of human safety on the roads. Common sense demands that lawyers should focus on the cause rather than the symptom. Promoting humane nonlethal methods of wildlife management can protect humans and nonhumans alike.

Federal conservation laws

The United States federal government, pursuant to the commerce clause or treaty power, has the authority to regulate, manage, and conserve wildlife concurrently with the states. There are a number of federal statutes and regulations that provide protection for certain wildlife, including endangered species,[102] migratory birds,[103] wild horses and burros,[104] and marine mammals.[105] These laws have a similar general structure. First, the law defines what animals are governed. Second, the law grants authority to a federal agency; for example, the Department of Agriculture and/or the Department of the Interior, Fish and Wildlife Service, to enact regulations that implement the goals of the statute. Such regulations limit conduct that will harm the covered animals but provide for permits to allow harmful conduct if an impact statement detailing the effects of the conduct on the animals demonstrates that the benefits to humans outweigh the harmful effects to the protected animals.

The terminology used to describe the purpose of the various laws governing wildlife is telling. The continuing property theme is clear through use of terms such as "preserve," "conserve," and "manage," each suggesting that animals are a resource for human use. Moreover, under these laws wildlife species are protected, not individual animals.

Perhaps the best example of these laws in the United States is the Endangered Species Act (ESA) originally enacted in 1973 and amended several times since. The ESA is deemed by many to be the most protective of the wildlife/environmental statutes and "uniquely draconian in its proscriptions [on human conduct] when compared to other federal environmental statutes."[106] The ESA is highly controversial because it limits human use and development of lands that are the natural habitats of protected species. The United States has a very strong commitment to the protection of private property rights.[107] Thus, laws that impose limits on owners' use of their property are politically controversial.

The Endangered Species Act

In the United States, the ESA protects threatened and endangered species from extinction through affirmative programs and conservation of their natural habitat.[108] Recall that in Chapter 3 we discussed a rather novel use of the ESA to attempt to protect the Ringling Brothers' Circus Asian elephants from abuse. Here we will discuss the more traditional use of the ESA – to protect threatened and endangered species in the wild.

An endangered species is defined as "any species which is in danger of extinction throughout all or a significant portion of its range ..."[109] A threatened species is defined as "any species which is likely to become an endangered species within the foreseeable future throughout all or a significant portion of its range."[110] The taking of a protected animal, defined as to "harm, pursue, hunt, shoot, wound, kill, trap, capture, or collect, or attempt to engage in any such conduct,"[111] is prohibited.[112] Any individual, group, or state agency may petition the Secretary of the Interior to list a species as endangered or threatened. The decision to list a species must be made solely on the basis of the best available scientific and commercial information, without accounting for economic implications.[113] "Undesirable" species, such as insect pests considered dangerous to humans, are exempted from protection.[114]

Once listed, the Secretary must designate critical habitat, defined as areas where the species is found or where resources essential to its conservation exist.

> Any area, whether or not federally owned, may be designated as critical habitat, but private land is only affected by critical habitat designation if some federal action (e.g., license, loan, permit) is also involved. Federal agencies must avoid "adverse modification" of critical habitat, either through their own actions or activities that are federally approved or funded.[115]

The Secretary may consider economic factors when designating critical habitat. Further, the Secretary is authorized to develop recovery plans for the species; however, the regulations provide few details and the plans are not binding on federal agencies or others.[116] The Secretary may also acquire land to conserve the species.

Federal agencies and state governors may apply for an exemption to allow an action that may harm a protected species to proceed. Approval of five members of a committee of six state and federal officials, commonly called the "God Squad," is needed to grant the exemption.[117] For non-federal actions,

> the Secretary may issue permits to allow "incidental take" of species ... The applicant for an incidental take permit must submit a habitat conservation plan (HCP) that shows: the likely impact; the steps to minimize and mitigate the impact; the funding for the mitigation; the alternatives that were considered and rejected; and any other measures that the Secretary may require.[118]

The ESA also implements the Convention on International Trade in Endangered Species of Wild Fauna and Flora (CITES) for the United States. Such provisions focus exclusively on controlling the import and export of living or dead organisms and their products that may harm a protected species.[119]

Some advocates claim that the ESA embodies "[t]he legal idea that a listed nonhuman resident of the United States is guaranteed, in a special sense, life and liberty."[120] Moreover, many lawmakers and citizens affected by the ESA claim that the statute "systematically subverts human needs to those of lesser creatures, ... [and] pit[s] the interests of wildlife against those of landowners."[121] However, as one scholar who studied the justifications for the ESA in depth has explained, while there are nature-centered justifications for the ESA, lawmakers enacted the ESA based primarily on human-centered justifications and enforcement of the ESA continues upon that theme.[122] Nature-centered justifications are based on a belief that other species have inherent interests, independent of human interests, and a right to exist in their own right. Human-centered justifications are grounded in the extrinsic and intrinsic benefits to humans of saving animal species. [123] The following is a summary of the arguments this scholar makes to support the claim that the ESA was enacted primarily to promote human-centered interests.

A review of the legislative history of the ESA suggests that Congress, in enacting the ESA, was most concerned with the instrumental effects of the law to benefit human interests by (1) protecting the earth's genetic resource, (2) maintaining the health of the nation's ecosystems for our own survival, and (3) securing aesthetic benefits.[124] The Congressional record supports this contention with statements like: "This legislation reflects our growing awareness that the continued existence of flora and fauna on our spaceship Earth have a direct bearing on the continued existence of man;"[125] "When we threaten endangered species, we tinker with our own futures;"[126] "Many of these animals simply give us [a] esthetic pleasure. We like to view them in zoos and in their natural habitats."[127]

There are also intrinsic human-centered reasons behind the ESA. Congress appeared to agree that humans, as the superior species, have a moral duty of stewardship over the planet, and that once a species is lost, it is lost forever. For example, Rep. Frank D'Annunzio stated:

> Irrespective of any instrumental benefit, many policymakers seemed to cringe at the thought of irrevocably erasing another form of life for all time. The additional fact that it has taken millions of years for the evolutionary process to create the species presently in existence seems to sharpen this sense of moral responsibility.[128]

Nature-centered justifications are harder to detect in the legislative history. Senator Cranston expressed a nature-centered justification for the ESA before the House Subcommittee in 1972, quoting Albert Schweitzer, and stating:

> The great fault of all ethics hitherto has been that they believed themselves to have to deal only with the relations of man to man. In reality, however, the question is what is his attitude to the world and all life that comes within his reach. A man is ethical only when life, as such, is sacred to him, that of plants and animals as that of his fellow man.[129]

Apparently, however, this view was not universally shared. No mention of such nature-centered justifications is found in the text of the ESA, the House or Senate Reports, or the Congressional debates, and no other member of Congress made any reference to such justifications as a primary rationale for the ESA.

In 1978, the federal exemption process was added to the ESA in direct response to the Supreme Court case *Tennessee Valley Authority* v. *Hill*.[130] In that case, the court ruled that "[t]he plain intent of Congress in enacting this statute was to halt and reverse the trend toward species extinction, whatever the cost," and halted the completion of a dam that was 80 percent complete and had already cost some $53 million in order to save from extinction the snail darter, a small fish native to East Tennessee who feeds primarily on aquatic snails.[131] This amendment to add such an exemption further supports Congressional intent to allow human economic interests to outweigh the inherent interests of other species. The committee may exempt any federal project when five members agree that the "project will confer more benefits to humanity than will preserving a certain species,"[132] even if the project may extinguish the species. The inquiry treats nonhuman species as resources that benefit humans. They represent merely one benefit to humans that are weighed against other human benefits. Further supporting the human-centered rationales for the ESA are the facts that certain "undesirable" species are exempt and that only species, not individual animals, are protected. The bottom line appears to be that the ESA treats animals as a human resource and protects them only when it is in the best interests of humans.

There are a few federal statutes that protect individual wildlife, such as the Fur Seal Act[133] (FSA) and the Bald and Golden Eagle Protection Act[134] (BGEPA). Both of these statutes prohibit the taking of an individual animal independent of the status of the species under the ESA. Are these laws examples of Congress protecting animals' inherent interests? Arguably not. "The FSA was enacted 'to conserve and manage the fur seal herd' in the face of international sealing pressures, thus ensuring a profitable 'annual harvest of sealskins' for the United States, Canada, Russia and Japan."[135] The BGEPA was enacted to protect human interests in the bald eagle as "a symbol of the American ideals of freedom."[136]

A case study: ESA remedies for endangered wildlife

Although the ESA is designed to protect species from extinction, the animals who comprise the protected species have no standing to enforce these laws. However, the ESA provides for a private right of action for concerned individual humans and organizations who meet the standing requirements to file suit to enforce the ESA. The remedy sought by the plaintiffs often is a permanent injunction, a court order restraining acts allegedly in violation of the ESA that will threaten the protected species. Moreover, not only do the plaintiffs seek a permanent injunction, one granted at the end of trial if the plaintiffs prove their case, they often seek a preliminary injunction to prevent harm to the animals while the case is litigated. Without such relief the case may be moot by the time the decision is final because the harm, the taking of the animals, may have already occurred. Whether a court will grant a preliminary injunction raises some very interesting and controversial issues. The following is a case study involving years of litigation to protect the grey wolf in the western United States from extinction, demonstrating some of these issues.

In 2006, several humane organizations filed suit against the Secretary of the Interior and the Director of the Fish and Wildlife Service (FWS), the entity that enforces the ESA.[137] The plaintiffs filed for injunctive and declaratory relief to prevent the killing of forty-three endangered gray wolves in Wisconsin on the basis that the permit FWS issued to Wisconsin to allow a lethal depredation control program was illegal. The plaintiffs immediately moved for a preliminary injunction to enjoin the lethal depredation control program until the case was decided.

A preliminary injunction is an extraordinary remedy because it is issued before there has been a full trial on the merits of the case. Nevertheless, if plaintiffs can demonstrate (1) a substantial likelihood of success on the merits, (2) irreparable injury if the preliminary injunction does not issue, (3) that the balance of harms favors plaintiff, and (4) that issuance of the preliminary injunction is in the public interest, a court will issue the injunction. In this case, the plaintiffs succeeded.

First, the court found that the plaintiffs were likely to succeed on the merits. The FWS issued the permit to allow the killing of 43 endangered gray wolves pursuant to Section 10(a)(1)(A) of the ESA. This section states that the Secretary of the Interior may permit the taking of an endangered animal "for scientific purposes or to enhance the propagation or survival of the affected species."[138] The regulations implementing this section require that the Secretary, in issuing a permit, take the following factors into account: (1) whether the project's purpose justifies the take; (2) whether the applicant has sufficient expertise and resources to accomplish the purpose; (3) the probable effects, direct and indirect, on the species; (4) whether the project would conflict with other programs intended to enhance survival of the species; (5) whether the purpose would likely reduce the threat of extinction of the species; and (6) the opinions of experts.[139]

The plaintiffs challenged the issuance of the permit on the basis that the lethal depredation program is contrary to Section 10(a)(1)(A) itself, making consideration of these factors unnecessary. The court agreed. Wisconsin's reason for conducting a wolf lethal depredation control program was that "in the absence of adequate measures to control known depredating wolves, public support for wolf recovery and wolf reintroduction programs will likely erode and individuals will resort to illegal killing to protect their pets and livelihood."[140] In other words, the purpose is to "increase social tolerance" for the gray wolf. The plaintiffs argued that such a purpose contravenes the express language and purpose behind the issuance of a permit "for scientific purposes or to enhance the propagation or survival of the affected species." The court stated:

> The language "propagation or survival of the affected species," is, on its face, antithetical to the killing of 43 members of an endangered species barring some direct and immediate danger imposed by the individual animals killed to other members of the species. While the Court can contemplate one circumstance in particular where a lethal take might be permitted to enhance the propagation or survival of the species – where an individual wolf had mange or some other communicable disease that could ultimately result in the death of other *wolves* – it is counterintuitive to authorize the killing of *endangered* animals rather than to authorize some non-lethal method of "take" to enhance their propagation or survival.[141]

Wisconsin argued that the program would enhance wolf propagation because without the program to increase social tolerance the "intolerant stakeholders … in the absence of an effective government-sanctioned wolf control program will engage in unregulated and unsupervised takings to protect their pets and livelihood."[142] Such takings would result in more killing than under the depredation program. The court described this argument as applying "a labyrinthian analysis that does not comply with the text of the statute on its face."[143] In fact, the court agreed with the judge, who had heard a similar argument in a prior case, and stated: "I am baffled by the government's position here. I have to be perfectly frank. I have a hard time understanding the notion you kill the wolves to save the wolves."[144]

Second, the court held that even a finite number of animal deaths of an endangered species constitute irreparable injury. The defendants argued that unless the number of deaths rose to the level that would jeopardize the entire species, the harm was not "irreparable." The court claimed that such a standard "would stand the ESA on its head. Without the ability to enjoin illegal taking under the ESA, courts would be without power to prevent harm to endangered species before a species was on the brink of extinction."[145] The court further stated that unlike a situation where the deaths are uncertain, this is a case where the deaths will occur.

> Environmental injury, by its nature, can seldom be adequately remedied by money damages and is often permanent or at least of long duration, *i.e.*, irreparable. If such injury is sufficiently likely, therefore,

the balance of harms will usually favor the issuance of an injunction to protect the environment.[146]

Third, in balancing the harms, the court noted that Congress, by enacting the ESA, had indicated a balance that heavily favors plaintiffs seeking to protect an endangered species. Moreover, the taking of wolves pursuant to this permit will not only "lead to the deaths of individual animals" but will also "disturb and stress intra-pack relationships, and cause a decline in the area wolf population" more broadly.[147] The state argued that the growth in numbers of wolves in their state had increased rapidly and as a result they currently occupy all areas of suitable habitat such that there are no areas left to relocate the wolves. The wolves are a nuisance to residents and a threat to livestock or other game that residents hunt. Thus, the only solution is a lethal one. The court told the defendants that they "seemed to miss the point." These wolves are first and foremost an endangered species, not a nuisance or a competitor to their sport hunters, and thus clearly the balance favored plaintiffs.

Finally, the public interest is promoted by the issuance of the injunction since an "examination of the language, history, and structure of the [ESA] ... indicates beyond a doubt that Congress intended endangered species to be afforded the highest of priorities."[148] Nevertheless, Wisconsin argued that "the public interest in the long-term health and recovery of the gray wolf population in Wisconsin will be best served by permitting the states to continue their depredation control activities" as this will increase public support of the wolves.[149] However, the court found this argument disingenuous, particularly in light of the alleged rapid growth of the gray wolf population in Wisconsin. The court concluded by stating:

> Congressional intent behind the adoption of the ESA and iterated throughout the language of the Act itself makes crystal clear that the 'public interest' lies in the protection of the endangered gray wolf-not in the lethal taking of 'problem' gray wolves in the hopes of creating a selected-for gray wolf population that never interferes with livestock or hunters' kills. Simply put, the recovery of the gray wolf is not supported by killing 43 gray wolves.[150]

Of note, not more than two years later, many of the same humane organizations filed a second suit to challenge the FWS's decision "to designate and delist a northern Rocky Mountain grey wolf distinct population segment under the Endangered Species Act."[151] They won again! The court issued the preliminary injunction finding both that the plaintiffs were likely to succeed on the merits and irreparable harm would befall the wolves during the pendency of the lawsuit.

Specifically, the court found that

(1) the Fish and Wildlife Service acted arbitrarily in delisting the wolf despite a lack of evidence of genetic exchange between subpopulations; and (2) it acted arbitrarily and capriciously when it approved Wyoming's 2007 plan to take gray wolves despite the State's failure to commit to managing for 15 breeding pairs and the plan's malleable trophy game area. In both instances, the Fish and Wildlife Service altered its earlier position without providing a reasoned decision for the change based on identified new information.[152]

Moreover, irreparable harm is likely since

more wolves will be killed under state management than were killed when ESA protections were in place. Idaho, Montana, and Wyoming each have public wolf hunts scheduled for th[e] fall ... and the states' defense of property laws permit the killing of wolves in more circumstances than defense of property regulations under the ESA. [Moreover], the killing of wolves during the pendency of this lawsuit will further reduce opportunities for genetic exchange among subpopulations.[153]

Interestingly, this court found it unnecessary to consider the last two factors of the preliminary injunction criteria – balancing of harms and public interest – because Congress, under the ESA, made it "abundantly clear that the balance [of hardships] has been struck in favor of affording endangered species the highest of priorities," and courts "may not use equity's scales to strike a different balance."[154]

5
Animals, the Constitution and Private Law

A constitution, whether written or unwritten, defines the basic rights and obligations as between the government and the individual human citizen, while private law defines the rights and obligations among human persons. Thus, these two sources of law are to protect human interests directly. As a result, animals, if present, almost always are present as objects – the property of their human owners or part of the "common" held for the benefit of humans – rather than subjects – referenced in their own right. To date, these sources of law provide little opportunity for robust protection of animals' interests, with a few notable exceptions. However, we will see that the manner in which these two sources of law are addressing animals has begun to change. Both the constitutional law and the private law are beginning to recognize the inherent interests of animals as sentient beings, and with time may do more to promote the animals' own individual interests independent of their owners.

Animals and the constitution

A constitution defines the fundamental values of a society, yet animals are absent from most constitutional texts. This suggests that their interests are not deemed of fundamental importance to the people of that society. The United States Constitution is an example of this. The preamble of the United States Constitution states:

> We the *People* of the United States, in Order to form a more perfect Union, establish Justice, insure domestic Tranquility, provide for the common defense, promote the general Welfare, and secure the Blessings of Liberty to ourselves and our Posterity, do ordain and establish this Constitution for the United States of America.

The *People* of the United States are protecting and promoting the interests of themselves and their posterity and no where within the text are entities other than persons mentioned. The United States Constitution is a relatively short document that outlines fundamental negative human rights, protecting the enumerated rights of the people by limiting the power of the government to infringe those rights.

Animals are not deemed persons under the U.S. Constitution.[1] Animals are classified as property, and are therefore present only indirectly in the Constitution and Bill of Rights to the extent that they are the property of a human. For example, the Fourth Amendment protects persons from unreasonable search and seizure of their property. The Fifth Amendment restrains the government from taking their property without due process of law or for public use without just compensation. We have discussed cases where owners have challenged animal control laws that limit their ownership interests in their animals as infringing their right to due process.

The Constitution also protects an owner's interest in an animal by prohibiting unreasonable search and seizure. A typical scenario involves an owner challenging a police officer's shooting and killing of her dog while searching her home. Unfortunately, there are numerous cases of police officers shooting and killing an owner's dog even though the dog poses no apparent threat. The owner alleges that the officer used unreasonable force when shooting the dog. Courts have held that when a police officer unreasonably kills a dog under such circumstances, the owner's rights to reasonable search and seizure are violated.

Consider the case of Bubba, a seven-year-old Labrador retriever/springer spaniel mix.[2] Bubba, along with his owner, Virginia, and other family members, were relaxing in their backyard when six police officers arrived at their home. The police had received an anonymous tip that a wanted felon had entered the home with a pit bull. One officer, Carter, brought a shotgun and later testified that "the best weapon for a dog is a shotgun through my experience." Bubba heard the police and ran towards the officers. Although the officers testified that Bubba was growling and exposing his teeth and gums, a neighbor witness testified he was coming out to greet the officers. Carter shot at Bubba twice, hitting him in the leg. Bubba limped into the bushes where he hid for ten minutes while Carter kept watch over him and prevented Virginia from retrieving Bubba or calling a veterinarian. When Officer Eyre arrived, he approached the bush in which Bubba was hiding which caused Bubba to emerge to go to Virginia. Undeterred by the growing number of neighbors yelling at the officers that Bubba was a good dog and not to shoot, Eyre's raised his handgun to shoot but then ordered Carter to shoot Bubba instead. Carter shot Bubba two more times, killing him. The alleged suspect and pit bull were not at Virginia's home.

Virginia filed suit against the officers for their unreasonable use of force that killed Bubba. The officers sought dismissal of the action on summary judgment based on the defense of qualified immunity. Qualified immunity is a defense for police officers and other state actors who reasonably believe they are acting lawfully. Government officials have qualified immunity for liability for civil damages if their actions did not violate "clearly established statutory or constitutional rights of which a reasonable person would have known."[3] Wisconsin law states that "the use of deadly force against a household pet is reasonable only if the pet poses an immediate danger and the use of force is unavoidable." The court denied the officer's motion as there were competing versions of the facts sufficient to go to trial. If Virginia could prove her version of the events at trial the officers would be liable for killing Bubba through the use of unreasonable force. In December 2008, the jury returned a verdict for the police officers.[4] No explanation accompanies a jury's verdict; thus, there is no way to determine why the jury found for the police. In any event, human owners and their companion animals are better protected if officers are held liable for unreasonably killing companion animals. Such liability creates an incentive for police departments to train officers on the proper way to handle dogs in such situations. As the Wisconsin law indicates, lethal force in these situations should be allowed *only* as a last resort when necessary to protect public safety.

In contrast to the U.S. constitution, some states within the United States and some countries expressly include animals within their constitutions. State constitutions grant the government power to manage and conserve wildlife and to regulate the treatment of domestic animals. Countries that include animals within their constitutions grant basic *positive* rights to the people. Such positive rights require the government to aspire to provide fundamental human necessities such as food and health care, and for the government and/or its citizens to protect and preserve the environment. Animals generally are present in these positive provisions protecting and preserving the environment.

Noted international animal law scholar, Dr. Gieri Bolliger, has studied the inclusion of animal welfare in European constitutions.[5] He notes that since a constitution embodies the overall values of a nation, the inclusion of animal welfare in the constitution reflects a commitment by the people of that nation to respect animals and their inherent worth as sentient beings. Protection of their interests is of the highest order. He explains that including animals at this level

> leads to an 'equality of weapons' in which the privileges of science, art, and religion or the freedom to choose a profession do not have priority over animal welfare concerns. Indeed, in any conflict between different constitutional rights, interests must always be weighed up in a balanced way. This means, for instance, that animal management must be adapted to the needs of animals and not depend solely on the economic interests of those who use animals.
>
> Constitutional anchoring ultimately helps to achieve better legal protection and the enforcement of animal welfare rights by enabling the comprehensive monitoring of legality and the development of independent practice in the use and interpretation of animal welfare regulations by administrative authorities and the courts.[6]

The following will survey a few examples of constitutional texts that include animals. How does such inclusion lead to an "equality of weapons?"

State constitutions

In the United States, the states have primary authority to manage their natural resources, including wildlife, as wild animals are deemed a resource for human use. The Florida Constitution, for example, creates a Fish and Wildlife conservation commission to "exercise the regulatory and executive powers of the state with respect to wild animal life, ... fresh water aquatic life, ... and marine life"[7] and regulates conduct affecting animals. For example, Article X, Section 16(a), states:

> The marine resources of the State of Florida belong to all of the people of the state and should be conserved and managed for the benefit of the state, its people, and future generations. To this end the people hereby enact limitations on marine net fishing in Florida waters to protect saltwater finfish, shellfish, and other marine animals from unnecessary killing, overfishing and waste.

Although animals are present in this environmental provision, focus on the "unnecessary killing, overfishing and waste" is designed to conserve marine animals as resources for human use rather than to protect marine life from pain and suffering for their own benefit. Moreover, the use of nets for scientific research or governmental purposes is exempt, further indicating that the law, while beneficial to the animals indirectly, is enacted with human citizens as the primary beneficiaries rather than the animals themselves.

In recent years, animal advocates have been somewhat successful in using the voter initiative procedures to amend state constitutions and add protections for the animals themselves. For example, the Florida Constitution, Article X, Section 21, states:

> Inhumane treatment of animals is a concern of Florida citizens. To prevent cruelty to certain animals ..., the people of Florida hereby limit the cruel and inhumane confinement of pigs during pregnancy as provided herein. (A) It shall be unlawful for any person to confine a pig during pregnancy in an enclosure, or to tether a pig during pregnancy, on a farm in such a way that she is prevented from turning around freely.

This provision clearly protects the animals' interests in being free from inhumane treatment. While the legal relevance of this specific provision as constitutional as opposed to statutory is not great, as a constitutional provision it cannot be overturned by the legislature but only by a second voter initiative. Thus, the state constitution can provide opportunities for animal protection lawyers to enact popular laws that cut against big business interests, that are protected from the legislative will, and that demonstrate the importance of protecting animals' interests by including them in the supreme law of the land.

Constitutions worldwide

In contrast to the United States, several countries include animals within their constitutions in a variety of ways. Because each country's legal structure and legal system are different, a facial comparison is not appropriate and an in-depth comparative analysis is beyond the scope of this introductory text. The purpose here is to identify a few of the countries that include animals in their constitutions and exemplify how they are included. Although the practical significance of these provisions depends upon implementation and enforcement by the sovereign, the inclusion demonstrates societal interest in protecting animals.

Most commonly, animals are included in the environmental provisions of constitutions. The provisions' primary purpose is to protect the environmental health of the country for the benefit of its human citizens. However, animals' inherent interests may also be included directly. For example, enacted in 1947, Article 51-A(g) of the Indian Constitution states:

It shall be the fundamental duty of every citizen of India to protect and improve the Natural Environment including forests, rivers, and wildlife, and to have compassion for all living creatures.

Note that while this provision is focused on environmental concerns broadly, a duty of "compassion for all living creatures" is also included.

The Brazilian Constitution is similar in this regard. In 1988, Brazil adopted a democratic constitution. Chapter VI, Section 225, Environmental Protection, reads:

All persons are entitled to an ecologically balanced environment, which is an asset for the people's common use and is essential to healthy life, it being the duty of the Government and of the community to defend and preserve it for present and future generations. (1) In order to ensure the effectiveness of this right, it is incumbent upon the Government to: ...

VII. *protect the fauna* and the flora, all practices which jeopardize their ecological function, *cause the extinction of species or subject animals to cruelty* being forbidden according to the law.

In 2008, Ecuador amended its Constitution to protect its ecosystems, including wildlife, from destruction from exploitive companies by adding the Rights for Nature Chapter.

Nature or Pachamam (a Quechua word meaning Mother Earth), where life is reproduced and exists, has the right to exist, persist, maintain and regenerate its vital cycles, structure, functions and its processes in evolution ... Every person, people, community or nationality will be able to demand the recognition of rights for nature before the public organisms.[8]

This provision signals a compassion for Nature and all living things, including animals, and grants standing to enforce the provision.

For over 100 years Switzerland has granted

constitutional protection to the "dignity of creature," explicitly according esteem to all non-human living beings, namely animals, at the highest legal level. The principle encompasses all legal aspects of human/animal interrelations and is supposed to restrict in particular the kind of treatment of animals that, although not necessarily associated with pain, suffering or damage, nevertheless affects other animal interests that must be respected by humans. Central features in this regard are the protection of animals from humiliation, from excessive instrumentalisation, and from intervention in their appearance.[9]

The European Union included animals in the Lisbon Treaty which amended the Treaty on the European Union and the Treaty establishing the European Community. The provision states:

In formulating and implementing the Union's agriculture, fisheries, transport, internal market, research and technological development and space policies, the Union and the Member States *shall, since animals are sentient beings, pay full regard to the welfare requirements of animals*, while respecting the legislative or administrative provisions and customs of the Member States relating in particular to religious rites, cultural traditions and regional heritage.[10]

The ratification of the Lisbon Treaty is likely to have a significant effect on the legal protection of animals throughout the European Union as the supreme text governing all 27 member states specifically recognizes the sentience and inherent worth of all animals, not merely their benefit to humans as a natural resource. Moreover, it requires that the member states "pay full regard" to animal welfare when enacting laws. Of course, the states retain the ability to respect their religious and cultural traditions and heritage. Although it is too soon to know the effect this provision will ultimately have, no other constitutional-like[11] text to date has affirmatively defined animals as sentient beings, promoted their inherent interests independent of human interests, and directed that the law protect them.

A case study: animal protection in the German constitution

The German Constitution received substantial media attention when, in 2002, Article 20a was amended to include "and the animals." Article 20a now states:

The state, aware of its responsibility for present and future generations, shall protect the natural resources of life and the animals within the framework of the constitutional order through the legislature and, in accordance with the law and principles of justice, the executive and judiciary.

This amendment was largely in response to the 2002 ritual slaughter decision of the Federal Constitutional Court.[12] The slaughter decision had held that a Muslim[13] butcher's constitutional rights to practice his religion and his occupation were infringed when he was denied a permit to perform ritual slaughter. The German Animal Protection Act prohibits the slaughter of warm-blooded animals without prior stunning except under limited circumstances, including when "granted an exemption slaughter ... where necessary to meet the requirements of members of religious communities ... whose mandatory rules require ritual slaughter [or] prohibit consumption of meat of animals not slaughtered in this way."[14] Although the butcher sought an extension of his permit under the exemption, it was denied. The court noted that constitutional law takes precedence over statutory law, and thus the German Animal Protection Act must yield to the butcher's constitutional rights and the butcher must receive an exemption to its provisions.

To those who wanted animal protection to be given greater weight, and not yield per se to constitutional claims, the solution was to include animal protection within the constitution. For many years, advocates had been seeking such inclusion, but it took this decision to persuade the government to act. Within one year, it was accomplished. The motivation for the amendment can be found in the explanatory remarks. The remarks state that humans have a moral obligation to respect animals and protect them from unnecessary suffering as they are sentient beings.[15] The inclusion of animals in the constitution is designed to demonstrate the importance of their protection and to increase the effectiveness of the German Animal Protection Act.[16]

However, it remains unclear what effect Article 20a will have on animal protection law in Germany, generally, and on the enforcement of the German Animal Protection Act, specifically. Media statements have ranged from claims that the provision is purely "symbolic and meaningless" to claims that "animals now dominate the German Constitution."[17] In reality, neither is accurate, and it is too soon to know for sure. One commentator concluded: "The major accomplishment of the amendment is in allowing national law to reflect the status that animals have attained in German society, and to bring current regulation to the level at which it was intended to function."[18] Thus, the expectation of this commentator and others is that this amendment will allow for the proper enforcement of the existing animal protection laws even if the law might arguably infringe a human constitutional right.

The German Constitution is quite different in substance and structure from the United States Constitution. The German Constitution includes both grants of fundamental rights, like the right to "the undisturbed practice of religion," enforceable by citizens, and statements of state objectives, like Article 20a. Although a state objective does not define an individual right directly enforceable, it is more than a mere declaration of policy. A state objective is a goal of constitutional magnitude such that the three branches of government are obligated to take the goal into account and to promote it to the extent possible.[19] The specificity of the objective dictates the nature of the obligation and the degree of government discretion. Thus, for example, the state objective of German unification – that East and West Germany become one – defined a very explicit outcome which ultimately was accomplished in 1990.[20] In comparison, Article 20a is rather vague in its objective and therefore grants more latitude to the government in deciding the ultimate goal as well as how to achieve it. Thus, since the provision does not allow standing for enforcement directly and is vague in its stated objective, its effectiveness is unclear.

Nevertheless, fundamental rights and state objectives have the same constitutional force. Thus, when in conflict, one does not trump the other per se. The courts and administrative agencies must balance the competing constitutional interests on a case-by-case basis and attempt to maximize each within the constraints presented.[21] This was the original purpose behind the amendment, to raise the protection of animals to constitutional dimensions and place it on equal footing with other constitutional rights, like freedom of religious practice and occupation.

Interestingly, the ritual slaughter case that triggered the amendment was on remand to the state administrative court when Article 20a was amended. The state administrative court reexamined the Federal Constitutional Court's decision in light of the constitutional amendment and held that the butcher must prove that his religion prohibits the consumption of meat slaughtered with prior stunning to a higher level than was previously necessary.[22] The federal administrative court disagreed and retained the original level of proof.[23] The court explained that the amendment changed the analysis of the constitutionality of the exemption provision, not the burden of proof. Previously, the question was whether not granting an exemption to the butcher was a violation of his fundamental rights. Now the court also must determine whether granting an exemption is compatible with the constitutional interest in animal protection. The court held that it is up to the legislature to reconcile these interests and, given that the Act provides for a religious exemption and had not been amended, the result is unchanged. This decision has been criticized by commentators suggesting that the court failed to appreciate the now heightened level of government interest in animal protection as it continued to refer to it as an important public interest, which was the same categorization it used before Article 20a was amended.[24]

This is the only case to date that has been decided under amended Article 20a. In this instance, it appeared to have no effect at all. It will be interesting to see whether the German Animal Protection Act can and will be amended to remove the religious exemption on animal slaughter in order to promote greater animal protection without being found in violation of the fundamental right to practice religion. Only time will tell whether the German courts will give greater weight to animal protection when balanced against other constitutional interests or be more liberal in their interpretation of the German Animal Protection Act. Further, it remains to be seen whether the legislature will seize the opportunity to strengthen the German Animal Protection Act in order to comply with their new constitutional objective.

Animals and private law

Regulating interpersonal relationships

Private law is designed to regulate interpersonal relationships and provide a remedy to those harmed by the wrongful acts of others. As discussed in Chapter 1, private law is divided among a number of substantive law areas including torts, contracts, property, family, and trust law, all of which can involve and affect animals. Within each of these areas an animal may be involved as an object – the property of a human – but rarely, if ever, as a subject – a recognized individual in his or her own right. For example, the law of property defines the ownership rights a human owner has over his or her animal and provides a cause of action for the human owner to recover the animal when another wrongfully takes him or her. Under tort law, the law of negligence creates standards of care that a person owes to the owner of an animal and provides a cause of action for compensation to the owner against one who breaches the standard of care and injures or kills the owner's animal. Moreover, animals may be the object of a contract. For example, people may enter into contracts to board, train, ride, groom, or purchase an animal. The law of contracts defines the requirements for the creation of an enforceable contract, when the contract is breached, and the remedy available if breached. In each of these cases the human seeks a remedy to compensate him/her for his/her own injury as a result of his/her loss of, or harm to, his/her animal. In order to compensate the owner for his/her loss, the court must determine the value of the animal to the owner. The valuation of the animal to the owner, as we will see, has changed over time and is an area in which the private law is beginning to recognize animals' inherent qualities.

Family law has begun to address companion animals as subjects in limited circumstances. A common issue presented in a divorce proceeding concerns which partner in a marriage will gain custody of the companion animal upon the dissolution of the marriage. Traditionally, the court views the companion animal as an item of communal property and determines ownership based upon traditional property rules.[25] However, slowly more progressive courts have begun to recognize that companion animals are sentient beings, considered members of the family, and thus uniquely different from any other form of communal property, such as the house, car, or stereo. As such, a few trial courts have begun to consider the interests of the companion animal in awarding custody.[26]

Finally, under the law of trusts, many states now provide for pet trusts.[27] A pet trust allows the owner to plan for the care of his/her companion animal upon the owner's incapacity or death by putting money into a trust account and declaring who shall care for the companion animal, along with the caretaker's responsibilities to the companion animal. In these cases, the law is designed to benefit the human owner while also addressing the animal's interests.

Protecting the owner's interests

A primary goal of private law is remedial. The harmed plaintiff seeks a remedy from the defendant who caused the harm. Animals cannot file suit on their own behalf. Moreover, even if owners request a remedy on the animal's behalf, the court will not recognize their injury independent of the harm to the owner. For example, in *Oberschlake* v. *Veterinary Assoc. Animal Hospital*,[28] the owners of a dog named Poopi took her to the vet to have her teeth cleaned. While under anesthesia, the vet negligently attempted to spay her although she had already been spayed as a puppy. Poopi's owners sought recovery for Poopi's own emotional distress of being "wrongfully spayed." The court stated succinctly that

> [a]lthough Poopi was obviously directly involved in the incident, a dog cannot recover for emotional distress – or indeed for any other direct claims of which we are aware. We recognize that animals can and do

suffer pain or distress, but the evidentiary problems with such issues are obvious. As a result, the claims on Poopi's behalf were ... not viable.[29]

Unless or until private law recognizes animals as independent beings who are owed a duty of care directly and have standing to enforce that duty, private law is unable to protect the animal as subject rather than a mere object. Interestingly, the court tried to further justify the decision based on the "evidentiary problems" posed by allowing Poopi to recover for her injury. What are the evidentiary problems? Obviously, Poopi cannot testify intelligibly to humans, which in itself creates a difficulty in valuing her injury. However, the law generally values harm on an objective basis, and thus the subjective testimony of the individual harmed is only one piece of evidence. As humans, although we cannot know the full extent of a dog's pain and suffering, we do know that dogs feel pain and suffer. Thus, the courts could develop a mechanism for assigning a reasonable objective value to a dog's pain and suffering.

The law, however, does allow recovery to the human owner for the loss of his or her animal. One element common to the private law areas of property, tort, or contract, and informs societal views of animals, is the element of damages. If an animal is the object taken or harmed or the object of a breached contract, the same element is present – the compensation due to the person with ownership in the animal. How the law values the animal in determining the appropriate damages, or compensation due the owner, is an important indicator of the laws' recognition of the animals' worth, at least to its owner.

A case study: valuing companion animals

Common law

The classification of animals as personal property dictates the remedies available for harm to or loss of an owned animal. Moreover, the value is a function of the human's use of the animal. Animals owned as "commodities" for economic gain (like those used in farming for example) do not present as serious a problem of valuation as companion animals. This is so because, traditionally, courts have been reluctant to recognize the value of an animal beyond the "fair market value" of the individual animal. Increasingly, however, as humans' relationships with their companion animals have changed and their companion animals have become more like family members, courts are recognizing that their value to their owners may include non-economic damages, damages to compensate for the inherent worth of the animal to the owner.

Available remedies for owners of companion animals are of great import to lawyers and owners alike. First, a proper remedy is essential to offsetting the costs of litigation in order to make a lawsuit a viable option. Even if an owner does not care about the remedy and wants to file suit against someone who has wrongfully injured or killed the owner's companion animal as a matter of principle, most owners do not have the resources to bring suit if the available monetary remedy will not at the very least exceed the cost of the lawsuit itself. Unless the animal is an economically valuable race horse or show dog, for example, the fair market value of the animal is negligible compared to the costs of suit, and therefore few lawsuits are filed. Second, for the system to achieve its remedial goal, the remedy must properly reflect the value of the animal to its owner. As courts recognize that the true value of a companion animal includes companionship value to the owner, owners will be more accurately compensated for their loss, more suits will be brought to enforce the law, providing greater protection to the animals and their owners.

Historically, courts only viewed animals used by their owners for commercial purposes as valuable – for example, animals used for food or entertainment – based on their economic worth.[30] Since the average dog or cat kept as a companion animal has no real economic worth, the courts did not recognize the loss of a companion animal as harm to the owner. Over the years, courts began to recognize that even companion animals have some economic value. Under traditional property theory, compensation for the loss of a companion animal is limited to fair market value or replacement value, if any, and "some special or pecuniary [economic] value to the owner, that may be ascertained by reference to the usefulness and services of the dog."[31] If the animal is injured, the recovery is the difference between the fair market value before and after the injury or the cost to repair (veterinary costs). However, many courts limit the cost to repair to the fair market value of the animal.

Additionally, the owner may be compensated for consequential losses incurred; for example, lost income likely to have been received by the owner from the animal. While it is more difficult to establish potential future earnings from the animal, evidence of past income received and demonstration of the likelihood of continued income may suffice. In the case of animals used for breeding, courts generally refuse to compensate for the loss of future progeny as too speculative, but may include the animal's breeding potential in the valuation of the lost animal. Other losses that may be recoverable include reimbursement for money spent on the animal for training, microchipping, or other similar expense.

For example, in *Bueckner* v. *Hamel*,[32] the defendant shot a Dalmatian and Australian shepherd owned by the Hamels. Both dogs were relatively young – two and three years old, respectively – purebred, and registered with reputable clubs. The Hamels introduced the following evidence: (1) they had purchased the dogs with the intent to breed them, (2) they had already picked out a male for the Australian shepherd, (3) the dogs could be expected to breed once per year and produce six to eight puppies, and (4) the market price for the puppies. In this case the dogs had not yet been bred. The court allowed the introduction of the evidence to assess the value of the two dogs themselves, but disallowed recovery for their prospective progeny. In *Petco* v. *Schuster*,[33] Licorice, a 14-month-old miniature schnauzer, slipped her leash and ran from a Petco employee who had taken her out for a bathroom break during her grooming. She was ultimately killed in traffic. The owner recovered the fair market value of Licorice ($500), her expenses to send Licorice to training school ($892), and the costs of microchip implantation ($52.40).

For the typical companion animal, the fair market value is minimal and there are no consequential damages because the companion animal does not generate income. In fact, the value of the animal, recognized by the law, arguably could decrease with time as the animal ages, just like the value of some personal property depreciates over time. Thus, if a veterinarian, acting grossly negligent, kills an owner's ten-year old domestic shorthair cat whom he/she adopted from a shelter as a kitten, the owner's damages would be the market value of a ten-year-old cat in a shelter.[34] Does this value truly reflect the owner's loss? Most animal owners would think not. What they have lost is a companion, a member of the family whom they have loved and cared for over the years. Generally, courts refuse to include such non-economic damage as a component of the animal's "special or intrinsic value."[35] Recently, however, a few courts and legislatures have begun to allow for the award of non-economic damages for the loss of a companion animal.

As a legal matter, however, this is not all that unique. The common law recognizes that some items of personal property may have no market value and yet be very valuable to the owner. For example, family heirlooms and photo albums may have no value to anyone else, but the owner places great value on them – they are both irreplaceable and priceless. Thus, the common law carves out an exception to the market value rule when an item of personal property has no market value. In this situation, the owner may recover the inherent worth of the item to the owner. Issues of proof are difficult, but not insurmountable. The court will allow such recovery so long as the alleged compensatory amount is objectively reasonable. Another example is when a person wrongfully cuts down an owner's trees. In valuing the compensation to the owner, courts have recognized the intrinsic value of a tree, including an aesthetic value based on the appearance of the tree to the owner, and a utilitarian value based on the shade the tree provides the owner.[36] Note that the intrinsic or inherent value of the tree is determined from the perspective of the owner and his or her use of the tree. The same is true for loss of an animal; the inherent value, if recognized, is not the inherent value of the animal to him- or herself, but the inherent or intrinsic value of the animal to his/her owner.

Further, under traditional rules, the owner can often recover his/her lost utility in the property. For example, if an automobile is destroyed beyond repair, the owner may recover both the market value of the automobile as well as the loss of use of the automobile until it can reasonably be replaced. This "loss of use" is often calculated based on the cost to rent a car for the time that the owner is without his/her own automobile. In the case of a typical companion animal, the primary utility of the companion animal is to provide the owner "companionship." Thus, under traditional property law, the owner should be able to recover the loss of companionship resulting from the loss of his/her companion animal.[37] However, few courts have awarded loss of companionship, either because they refuse to recognize this utility of the companion animal or because they find that the companion animal has a market value, and thus intrinsic value – for example, loss of companionship – is not recoverable. In *Petco*, although recognizing all of these issues, the court, bound by Texas Supreme Court precedent, refused to award damages for Licorice's companionship value to her owner.[38]

It is important to distinguish loss of companionship value of the animal from the emotional distress damage the owner may suffer as a result of losing his/her companion animal. The claim for emotional distress injury is brought under a theory of intentional or negligent infliction of emotional distress. The emotional distress that the owner suffers is a function of the relationship the owner had with his/her companion animal – the closer the owner was to the companion animal, the more distress he/she would likely suffer. However, the distress the owner suffers is not a measure of the value of the live animal to the owner. Distress is a negative emotion suffered by the owner caused by the harm done to his/her companion animal. In contrast, companionship is a positive emotion that the companion animal, while alive, gives to the owner. Upon the loss of the companion animal, the owner loses the companionship the companion animal had provided.

Under traditional property theory, owners generally cannot receive emotional distress damage for the loss of property. The law *only* awards emotional distress damages when a person loses a special human loved one, like a spouse. Of course, a lost spouse is not the property of the living spouse seeking emotional distress damages, thus there is no corresponding lost value of the lost spouse to the survivor as there would be if the survivor lost his/her companion animal. However, under wrongful death statutes, the surviving spouse is able to receive the value of financial support and services that the lost spouse would have provided. This is arguably reflective of the value of the dead spouse to the survivor. Very few state statutes allow for emotional distress damages for loss of a companion animal, but it is highly controversial. Opponents argue that allowing emotional distress damages for the loss of an animal is an effort to equate an animal to a human, because emotional distress damages are only available under the law for the loss of a close human family member.

Arguably, allowing emotional distress damages for loss of companion animals, when such damages have been available only for the loss of a human loved one, extends the law beyond its current and traditional scope. While this may be supportable, and in fact proper, it is an extension of existing law. However, including the lost companionship as part of the value of the companion animal to the owner maintains the status of the animal as property of the human and is consistent with current doctrine. Valuing companionship merely recognizes the inherent value of this specific property, the companion animal, to the owner to more accurately reflect the owner's loss. Notably, whether the law allows emotional distress damages or compensates for loss of companionship, the benefit is to the human owner, not the companion animal. Nevertheless, as the law acknowledges the emotional relationship owners have with their companion animals, the more likely, with time, the law will acknowledge the inherent qualities of the animals themselves. This, in turn, could lead to better protection of animals' interests independent of their owners' interests.

Statutory law

In recent years, a few states have enacted statutes that expressly allow for damages for the loss of a companion animal, including non-economic damages. The enactment of these statutes demonstrates that changes in society and developments within the common law can lead to political and legal advances in the positive law. The basic elements of three state statutes – Connecticut,[39] Illinois,[40] and Tennessee[41] – are summarized below.

Definition of "pet": The statutes generally define what animal qualifies as a pet for purposes of recovery. The primary purpose is to limit recovery of damages to certain species and to those who own the animals for purposes of companionship rather than farming, research, or other commercial purpose. Thus, for example, Connecticut defines "companion animal" as "a domesticated dog or cat that is normally kept in or near the household of its owner or keeper and is dependent on a person for food, shelter, and veterinary care, but does not include a dog or cat kept for farming or biomedical research practices." Interestingly, Illinois allows recovery for any "animal" to which a person has a right of ownership, neither limiting the owner's utility of the animal to that of companionship nor the species to dog or cat.

Conduct establishing liability: States may hold a person liable for intentional or negligent conduct that harms an animal. Connecticut limits liability to "any person who intentionally kills or injures a companion animal." Similarly, Illinois limits liability to a person who engages in "an act of aggravated cruelty" as defined under the criminal code or killed by a person acting in bad faith. Under Tennessee's statute, a person may recover when their "pet … is killed or sustains injuries that result in death caused by the unlawful and intentional, or negligent, act of another or the animal of another." However, in the case of negligence, the animal must have suffered the death or injury while on the property of his or her owner or under the control and supervision of his or her owner.

Damages allowed: Each state allows economic and non-economic or punitive damages within certain limits. Economic damages allowed include "expenses for veterinary care, the fair monetary value of the companion animal, … burial expenses" and "other expenses incurred by the owner in rectifying the effects of the cruelty, pain and suffering of the animal." Connecticut allows recovery for punitive damages; Illinois allows recovery for the "emotional distress suffered by the owner;" and Tennessee compensates "for the loss of the reasonably expected society, companionship, love and affection of the pet," but limits such recovery to $5,000.

Exemptions: The state may also exempt certain actors from specific types of damages. Negligent veterinarians are exempt from liability for non-economic damages in Tennessee. Veterinarians in Connecticut are exempt from punitive damages and attorney's fees "while following accepted standards of practice of the profession." Moreover, state actors and animal-related non-profits are similarly exempt from liability for punitive damages and attorneys' fees.

These statutes codify certain substantive elements of the common law while modifying others, including allowing the recovery of non-economic damages. For example, although the common law will allow recovery of economic damages for the negligent killing of a companion animal, the statutes differ as to the level of culpability required to award non-economic damages. Illinois requires proof of bad faith; Connecticut requires intent; and Tennessee allows recovery for negligence, but only when the injury occurs on the owner's property or the companion animal is under the owner's control and supervision. The statutes also differ as to the type of non-economic damages allowed – Connecticut allows punitive, Illinois allows emotional distress, and Tennessee allows loss of companionship. Finally, the statutes provide exemptions for recovery of non-economic damages for certain actors. Such exemptions are not necessarily recognized under the common law, partly because the common law does not allow recovery of non-economic damages.

In sum, although private law currently remedies only harms to humans and treats animals as objects of property, one could argue that the recent movements (1) to allow non-economic damages for the loss of one's companion animal, (2) to consider the animal's interests in awarding custody, and (3) to grant animals the status of beneficiaries under trusts, reflect a growing compassion for and greater understanding of the sentience and inherent worth of animals. The law progresses slowly. Today, animals are still viewed as the property of a human and still lack independent standing to recover for their own injuries. However, with courts and legislatures beginning to compensate owners for the lost companionship that their companion animal provides, they are recognizing the inherent qualities of these animals as beings who share love and companionship with humans. This is the first step in appreciating their sentience and inherent worth as individuals. Further, if the true value of animals is recognized by the law, the law, in turn, will have a greater deterrent effect on human actors and fewer animals will be victims of harm by third parties. Finally, by considering the animal's interest in awarding custody, the courts are directly accounting for the animal's interests independent of the human's interest.

Of course, the only animals represented in the private law are those owned by humans. For example, free-roaming or feral cats currently lack any representation in private law since arguably no human has standing to recover for their loss. Should the private law recognize unowned animals? The animals could be represented in court by a human guardian ad litem or humane organization. But what would the remedy be if the defendant were found liable for harming the animal? Arguably, monetary damages could be placed in trust for the animal and the human caretaker, as trustee, could use the money to provide shelter and food for the animal. Of course, until the law recognizes the interests of the owned animal, independent of his or her owner, it is unlikely to recognize the interests of the unowned animal. Nevertheless, the progressive steps we have seen may lead to a greater recognition, appreciation, and protection of all animals' inherent interests under the law.

6
The Future: Animals as Subjects, an All-Inclusive Legal Regime

Welfare versus rights

Thus far, we have studied how animals fit within the existing legal regime. In this final chapter, we will look briefly to the future of animal protection law. What have we learned from a review of the existing law and how might the law be constructed to better protect animals? The focus of this chapter is to initiate a dialogue and to propose a few suggestions and examples of a legal regime that considers animals as subjects.

Changing the law to reflect animals as subjects and grant them equal consideration requires a change in the moral and ethical views of society. This fundamental question is a matter of philosophical theory. Do all sentient beings deserve to be treated as subjects under the law, to be given equal consideration, to be given rights? Philosophers have debated this question for centuries, and while there is no agreement, much progress has been made. There now is a growing consensus among philosophers for moral change in our consideration of animals. The law, in turn, is the means by which a society defines and enforces their philosophical views of what is morally and ethically required.

In the context of animals and their proper role within our human society, we might over-simplify matters greatly and say that philosophers who believe that animals deserve moral consideration generally fall into one of two camps. The first camp is often referred to as the "welfare" camp. Welfarists posit that animals have an interest to be free from pain and suffering, and humans, in turn, have a moral obligation to treat animals humanely. However, pure welfarists believe that it is morally and ethically proper for humans to use animals solely to serve human interests so long as humans provide for the animals' well-being. The second camp is often referred to as the "rights" camp. Philosophers in this camp argue that animals not only have an interest to be treated humanely, as the welfarists believe, but also have an interest in life itself. Thus, it is immoral for humans to use animals solely for human benefit even if the animal is treated humanely during such use. Within each camp there are a variety of approaches and nuances of position. However, for our purposes here, the relevant distinction is that between allowing or disallowing use of animals solely for human benefit.

As we have seen, the current law follows a welfarist approach. The law views animals as property, the object of human use. Moreover, the laws allegedly designed to protect animals provide standards for their treatment and welfare in order to protect them from *unnecessary* pain and suffering. Thus, the law, through the use of terms such as "unnecessary," provides for a utilitarian-type balancing where the human benefit derived from such use is balanced against the harm to the animal. Often, as we have seen, human interests trump.

Both philosophical camps argue that the current legal regime is speciesist and fails to properly protect animals' interests. Those in the welfarist camp argue that the law does not grant animals equal consideration. The law does not properly reflect the animals' interests in being free from pain and suffering, nor does it provide for the animals' interest in positive well-being beyond freedom from pain and suffering. Those in the rights camp argue that the law must fundamentally change such that humans are prohibited from using animals exclusively for human benefit and that animals' interest in their own well-being and life itself must be protected under the law. Proponents of a rights approach debate how best to achieve such a fundamental change in the law. Some believe that laws designed solely to improve animal welfare are a step in the right direction towards meeting the ultimate goal of achieving a legal right to life for animals. Others, like Gary Francione, believe that laws that are designed solely to improve animal welfare without also moving towards an eradication of legal use of animals prevent the legal regime from attaining the ultimate goal of granting animals a right to life. This is because pure welfare laws reinforce the notion that animals are property and may be used solely for human benefit. Philosophers and legal scholars in both camps have written extensively in this area.[1] Their writings demonstrate the rich variety and nuance of the philosophical approaches. While fascinating, a full discussion of the debate goes beyond the scope of this introductory legal text. Rather, this text will introduce some fundamental parameters of a legal regime that is non-speciesist, treats animals as subjects, and appreciates and protects their independent, individual, and inherent interests, and explore how the legal regime might differ under these two simplified approaches.

Problems for animals under current law

Moving towards a better legal future for animals requires understanding the deficiencies of the current system. What are the problems for animals under the current legal regime? Some of the problems go to the heart of the current regime, while others are less fundamental. Most fundamentally, as we have discussed throughout, the law characterizes animals as objects, property owned by humans. As currently constructed and interpreted, this characterization is the source of many of the deficiencies that exist. In order for the law to properly account for animals' interests, animals must be counted as subjects. This does not mean necessarily that their object status must be eradicated completely, but rather that subject status must be granted to them, at least in part.

Their characterization as objects also promotes the organization of the law around the human *use* of an animal – for example, the law governing companion animals, animals used for food, in research, or for entertainment. This organization not only reinforces the focus on the owner-object relationship between human and animal, but also results in different treatment of the identical animal depending upon the human use of the animal. Presumably, a given animal has the same inherent interests independent of his or her use by humans, yet the law does not acknowledge this. Thus, the legal protection provided a rabbit, for example, will be varied depending upon that rabbit's relationship to humans. The rabbit who is the companion to its human owner is governed by the state anti-cruelty laws and is protected from cruelty and neglect by its owner. The rabbit who is used for research or entertainment is governed under the Animal Welfare Act (AWA). As we have seen, the standards under the AWA are minimal, research requirements often justify the animals' pain and suffering, and enforcement is rare. The rabbit who is raised and slaughtered for food has no protection while on the farm, and the slaughter and transport laws often inadequately prevent pain and suffering of the animal. And the rabbit who lives in the wild can be hunted or trapped as a pest with little if any protection at all. Thus, the law treats this rabbit exclusively as an object of human use rather than as a sentient being with inherent interests that should be protected independent of the rabbit's relationship to humans.

Beyond this fundamental characterization, what other problems have we identified? The main problems may be divided into those that are structural and those that are substantive.

The current regime is structurally inadequate to properly protect sentient animals. First, animals have no voice in the legal system.

Animals themselves lack standing directly to file suit in a court of law to enforce the laws designed to protect them. Moreover, humans who seek to enforce the laws indirectly for animals have severe difficulties meeting the current standing requirements. Furthermore, animals lack representation within the political branches of the government, the legislature and executive branches. Second, the laws governing animals are generally rules of reason, not bright-line rules, using terms such as "unnecessary," "reasonable," or "to the extent possible." Such laws allow for too much discretion in interpretation and enforcement. Finally, the terminology used in the law to describe animals (for example, "pests," or "nuisance wildlife") and to describe government goals (for example, "conserve," "manage," or "control") often denigrates the animals and their interests.

The laws are also substantively inadequate. One main reason for this is that a primary motivation of all laws is to benefit humans. For example, a primary rationale for anti-cruelty laws is to promote a more civil and moral human society for human benefit. Thus, the anti-cruelty laws prevent individualized gratuitous violence against animals so that such violence is not in turn directed towards humans. However, as we have seen, the anti-cruelty laws in the United States generally exempt animals used for food or research as well as wildlife, or the laws exempt accepted agricultural practices, hunting, fishing, and trapping, allowing institutionalized abuse of these animals. The AWA, allegedly, is designed to protect the welfare of animals used in research, for example, but only so long as the research goals of the human researchers are not impaired. The private tort system is designed to compensate human owners for the loss of their property and fails to recognize the individual harm to the animal; and the wildlife conservation laws are designed to protect the ecosystem and environment for human use and pleasure and generally protect only species from extinction not individual animals from harm. Moreover, the governing standards are those generally consistent with the human owners' interest in having a healthy, productive, and useful animal, but when human interests conflict with the animals' interests, human interests trump.

Finally, laws that appear to have no impact on animals may indirectly affect their use and welfare. For example, the Farm Bill, previously discussed, was enacted to help sustain small farmers after the Great Depression by offering various price supports on a variety of crops. However, over time and with the advent of technology that transformed the farming industry, the Farm Bill was amended to provide huge government subsidies of corn and other grains. Such subsidies provided the economic incentive for animals to be fed corn rather than to graze in the open field. In turn, large farms absorbed smaller farms to the point where only a few now dominate the entire industry. Consequently, the animals were relocated in large numbers into factories, never to see the light of day, and to be treated cruelly. The tax system is another area where the law may indirectly affect animals and their well-being, given that the tax system is designed to incentivize certain activities and dis-incentivize others. As a practical matter, especially in the United States, a market-driven society, economic subsidies, and economic incentives play a large role in creating the environment in which all sentient beings live.

The goal of an animal protection lawyer is to improve the protection of animals under the law. Improvement of the law may be accomplished in a number of different ways and to a variety of degrees, however, "three things are required for improvement through regulation. First, there need to be laws [in place.] Second, there needs to be adequate enforcement, and third, there has to be compliance."[2] The second and third requirements are especially difficult to achieve given the limited resources and ineffective incentives we have discussed. However, using laws indirectly to create the economic incentives to provide for the well-being of animals and to require disclosure of information important to consumers, while educating the public through advocacy campaigns, may help change societal views and in turn increase support for laws that directly promote animals' interests while facilitating enforcement and compliance with such laws.

Let's focus first on the laws that directly affect animals. Proposed amendments to better protect animals fall along a continuum from minimal enhancement of the current legal regime to a complete legal recharacterization of animals and their relationship to humans. Leading legal scholars, falling in both philosophical camps, have proposed a variety of approaches along this continuum. Let's discuss four of the proposed approaches,[3] those of legal scholars – Cass Sunstein, David Favre, Steven Wise, and Gary Francione – and conclude with a fifth approach by Eleanor Evertsen and Wim de Kok, who drafted the General Animal Protection Act (GAPA) with the Coalition of Animal Welfare Organizations of the Netherlands. Note that the treatment of their approaches here is necessarily brief. The objective is to introduce the reader to the essence of their positions and to compare and contrast their ultimate goals and their methods for achieving those goals.

Approaches for improving the law for animals

Four approaches along the continuum

We begin with Cass Sunstein[4] who argues for maintaining the current regime with enhancements. Sunstein identifies several deficiencies under current United States law that, if corrected, would vastly improve animal protection while leaving the existing system intact. As a result, this approach is likely the most politically persuasive approach. The following are the issues and solutions he identifies. First, the law expressly exempts large numbers of animals from protection. For example, as we have discussed, there is no protection for farm animals used for food, and chickens, turkeys, and other domestic fowl (e.g. poultry) are exempted from the slaughter and transport laws. Moreover, approximately 95 percent of all animals used in research (birds, mice, and rats) are exempt from the protections of the AWA. These gaps must be closed by enacting laws to protect all animals from inhumane treatment on the farm, including birds within the existing laws, and expanding the coverage of the AWA to all animals used in research, including birds, mice, and rats.

Second, many of the welfare standards are inadequate and do not sufficiently alleviate animal pain and suffering. For example, the regulations under the AWA for dogs and primates are not adequate to meet their needs and should be improved to provide greater care and treatment to these animals. Third, enforcement is wholly inadequate. Most enforcement of the laws is by public entities, such as the United States Department of Agriculture (USDA), the Fish and Wildlife Service (FWS), the police, or prosecutors, and there are insufficient resources and incentives to enforce the laws. The solutions are to devote greater resources for public enforcement, to increase incentives for public enforcement, and to allow private enforcement of all laws by granting standing to animals, through representatives, to enforce the laws and alleviate some of the burden on public entities.

Next along the continuum is David Favre, who proposes that a new category of property, "living property,"[5] be created to better account for the needs and interests of domestic animals.[6] Favre explains that he is not an abolitionist and that "positive human communities can include animals that are owned/used by humans." Moreover, animals' status as "things" does not prevent their interests from being protected. In setting out his new paradigm of "living property," Favre expressly uses the term "ownership" rather than "guardianship" to describe the relationship between human and animal. He explains that nothing is inherently wrong with the concept of ownership – the concept may be beneficial when respectful or detrimental when oppressive. To emphasize the beneficial aspect, he defines an owner of an animal as "that human, or entity, who has responsibility for the animal in the context of the limited human and animal rights that will be set out." The following summarily describes Favre's new category of living property and the legal characteristics that attach.

Favre defines living property as "vertebrate animals who are property and shall be identified by either specific name or by group reference."[7] He then identifies the characteristics of an interest that qualify for protection.

> First, the interest has to be clearly knowable and articulatable. Second, the realization of the interest has to have the potential of being interfered with by humans or the government itself. Third, there has to be a remedy that can be implemented within our legal system. Finally, consideration of other points of public policy would have to be weighed in the balance.[8]

He explains that choosing among the interests that qualify is a judgment call based on a number of factors, including our understanding of the interest, potential conflicts with other interests, and the availability of a remedy.[9]

Next, Favre describes how including living property will affect three basic parameters of existing property law – the rights of the owners will be limited to some degree to accommodate the living properties' interests; the duties of human owners and non-owners towards living property will be imposed; and, living property will hold certain rights themselves.[10] Then, he defines the basic rights of humans in living property. These include title to living property, the ability to transfer title, and the ability to use living property. Each right must be carefully defined and circumscribed and the law must determine what uses are allowed and what living condition is appropriate for each legal use. Moreover, the duties humans owe to living property must account for the physical and mental well-being of the living property.[11] Finally, he sets out the basic contours of the rights of living property. At a minimum, these rights must include the right: (1) to have standing, (2) not to be held for or put to prohibited uses, (3) not to be harmed, (4) to be cared for, (5) to have living space, (6) to be properly owned, (7) to own property, (8) to enter into contracts, and (9) to file tort claims.[12]

Steven Wise offers yet a different approach: achieving animal rights "one step at a time."[13] Wise, as a practicing attorney and advocate for nonhuman animals, understands that progress is slow and impeded at every step by several obstacles. Wise identifies the obstacles as follows: physical, given there are at least 10 billion animals slaughtered for food annually in the United States; economic, given powerful industries rely on animal products; political, given humans, by definition, dominate the debate; religious/historical, given Judeo-Christian beliefs in the notion of the dominion (seen often as the domination) of humans over animals even if humans have basic obligations to nonhumans; legal, given the continued dichotomy between humans as persons and animals as property; and psychological/factual, given our ignorance of facts about animals' capabilities and emotions and reliance on inherent beliefs.[14]

To move the system forward, Wise proposes that every being with "practical autonomy" be entitled to basic liberty, relying on the "first principles of Western law: liberty and equality." Equality requires that we treat like beings alike. Liberty entitles a being to be treated in a manner consistent with and respectful of their individual characteristics. The irreducible minimum liberty rights of humans are bodily liberty (freedom from enslavement) and integrity (freedom from torture). He derives his notion of practical autonomy from two critical aspects of liberty: autonomy and self-determination. He defines practical autonomy as having the following capabilities: (1) desire, (2) intention to try to fulfill that desire, and (3) a sense of self-sufficiency that allows a being to understand they have wants and to attempt to attain them. Thus, "consciousness, but not necessarily self-consciousness, and sentience are implicit in practical autonomy."[15]

Wise then posits that practical autonomy is not necessary to be entitled to equality rights; however, animals with practical autonomy are entitled to basic liberty or dignity rights. Thus, basic liberty rights (in other words, personhood) should be given in proportion to the degree that a being has practical autonomy. He defines a scale of practical autonomy derived from the discipline of cognitive ethology, with 1.0 being absolute certainty that a being has practical autonomy. He then creates four categories of beings based on their practical autonomy scale value using the precautionary principle as a guide. The categories are: (1) 0.71 to 1.0, (2) 0.51 to 0.70, (3) at 0.5, and (4) below 0.5.[16] Under Wise's scheme, all beings in category 1 are presumed to have practical autonomy; those in category 4 are presumed not to have practical autonomy, those in category 3 are completely indeterminate, and those is category 2 must be treated under a proportionate scale. He admits that he is using a "human yardstick" on nonhuman animals, but explains that basic liberty rights are legally protected "because they are basic to human well-being, and the autonomy values we assign to nonhumans will be based upon human abilities and human values," which is a necessary bridge to a future when we might value nonhuman values as well.[17] Wise concludes that animals in category 1 and at least some in category 2, currently including great apes, Atlantic bottle-nosed dolphins, African elephants, and African grey parrots, measure up to human standards and deserve to be recognized as legal persons deserving of basic liberty rights.[18]

The fourth theorist is Gary Francione,[19] a leading proponent of the abolitionist view that the property classification of animals must be abolished and all sentient beings viewed as legal persons. Francione argues that it is only then that animals will have true legal "rights" – meaning the law recognizes and protects their inherent worth and forbids them to be used solely to satisfy human purposes. He explains that the full extent of their rights must be determined over time, but at a minimum they must have the right to live independently of humans. Francione explains that the only way to achieve this goal is to promote laws that are abolitionist; that is, that do more than merely improve animal welfare. Rather, they move towards the eradication of a property status for animals. He claims that laws that only improve animal welfare reinforce the notion that animals are property and do very little, in practice, to protect animals' interests.[20] Thus, for example, a law prohibiting the use of any leg-hold trap is preferred over a law that requires that trapping be done humanely.

Francione defines the attributes of a law that will allow for incremental change to achieve the goal of affording animals rights equivalent to those of persons as follows:

> The law must expressly prohibit an activity (as a per se or bright-line rule) and have the following attributes:
> (1) The prohibited activity must be essential to or constitutive of the exploitative institution.
> (2) The prohibition must recognize and respect non-institutional animal interests, the inherent interests the animal has independent of the institutional interests, and not simply ensure that the animal is used wisely to achieve a humane, institutional benefit.
> (3) Animal interests must be protected from interest-balancing by ensuring that they are not tradable independent of the justified benefit to humans.
> (4) The prohibition shall not substitute an alternative and supposedly more humane form of exploitation and shall be accompanied by a clear demand to abolish all institutionalized animal exploitation.

According to Francione, all of the stated attributes are essential to achieve personhood status for animals because otherwise the incremental change will reinforce the property paradigm that is sought to be abolished.[21] Thus, a law that bans the use of all animals in a particular type of experiment – for example, in psychological experiments – is abolitionist, according to Francione, because the animals have an interest in not being used for such experiments.[22] This law satisfies these criteria because it is (1) a prohibition banning all use of animals in a context that is essential to the research institution, (2) for the animals' independent interests, (3) where no interest-balancing or (4) substitute exploitation is allowed. He explains that banning the use of one species "might be acceptable if it were not based on a supposed moral superiority of that particular species and if no other species were proposed as an alternative."[23]

A progressive model act

The General Animal Protection Act (GAPA), written by Eleanor Evertsen and Wim de Kok, two scholars from the Netherlands, in coordination with the Coalition of Animal Welfare Organizations in the Netherlands (CDON) presents a progressive model for reform. [24] In 2008, the draft of a new animal law, entitled the Animals Act, was sent to the Netherlands' Parliament. The GAPA was written in response to this law as an alternative, based on the ethical principle that governs the European Union (EU) requiring EU law to acknowledge the intrinsic value of animals as living sentient beings. The GAPA defines the general parameters that the legislature must meet when enacting a law that affects the interests of animals in order to acknowledge their intrinsic value. The proposal addresses both individual instances of animal abuse as well as institutionalized uses of animals and imposes duties upon humans to protect the interests of animals in both contexts. The following is a description of this new paradigm and discusses how the provisions address many of the deficiencies of the current legal regime in the United States.

First, the GAPA governs "every live animal, irrespective of the species it belongs to, whether kept or living in the wild."[25] It then expressly defines fundamental terms that are used throughout to avoid subjective interpretations of what the terms mean. Critical to the paradigm is a definition of the three fundamental interests of all animals – welfare, health, and integrity. These interests include both physical and psychological conditions of the animals. Moreover, the definitions set the criteria well above a minimal standard. For example, the definition of animal welfare describes the animal living "in harmony ... with itself and its environment."[26] Animal health is defined as "the well-being of the animal

... not just the absence of illness or injury."[27] Integrity is defined as "the wholeness and sound condition of the animal and the state of living according to the ways and needs of its species, as well as the ability to sustain itself in a species specific environment."[28]

Next, they define the two competing interests to which the law speaks: the interest of the animals in their intrinsic value and the interests of humans in their use of animals. The intrinsic value of an animal is "the value an animal possesses and embodies as an individual being with its own life, its own experiences and feelings, simply because it is alive, regardless of any (added) value this animal may hold for humans."[29] Animal use is "any employment, exploitation or other form of use of animals for human purposes, interests or policies, including the management of nature reserves, execution of government policies and measures as well as the removal of animals from nature, in so far as it can be surmised that by these actions the recognition of the intrinsic value of an animal or animals is violated."[30]

The GAPA then sets the guiding principles and default parameters for all laws on the side that protects and promotes the interests of animals rather than those of humans. It accomplishes this by (1) requiring that "the intrinsic value of the animal can never by operation of law be made subordinate to the interests of man;"[31] (2) recognizing that animals, as sentient beings, with consciousness and feelings, "cannot be solely regarded as objects or commodities;"[32] (3) setting the standard for animal use at "no, unless" – no use of animals unless explicitly permitted in legislation;[33] (4) establishing the precautionary principle that when the impact of a use of an animal is not sufficiently clear, it is assumed the impact will be detrimental to the animal's welfare;[34] and (5) demanding that the law incorporate the "3Rs" with respect to all uses of animals, promoting the reduction, refinement and replacement of animals with existing alternatives while stimulating the use of alternatives.[35]

The law prohibits individual instances of animal abuse by prohibiting specific acts that infringe on animal welfare, health and integrity, and defines an obligation for all humans to provide animals in need with necessary care without delay.

The GAPA imposes a requirement that humans seek to improve animal's interests based on the most current scientific knowledge in animal welfare and health. The GAPA treats kept (for example, domestic) animals and wild animals differently. This distinction properly recognizes that the only relevant difference that affects humans' interaction with other animals is that humans live *with* kept animals and *among* wild animals. However, there is no fundamental distinction based on the variety of human uses of kept animals. For kept animals, their interests include sufficient staff who possess appropriate skill, knowledge and empathy and who provide sufficient quality of wholesome food, water and shelter as well as accommodation that (a) meet their social needs, such that they have same-species companionship when appropriate, space to groom, rest, move about, withdraw, and display their species specific behavior; and (b) stimulates play.[36] The guarantee for animals in the wild is their interest in "optimal freedom," defined as "a life undisturbed as possible in their natural habitat" or a sanctuary when necessary that meets the needs as described for kept animals.[37]

Finally, this proposed legislation addresses the existing problems of enforcement and review by requiring that (1) the parameters of the law be sufficiently clear and objective (per se rules); (2) the legal status as kept or wild be explicitly defined and, if kept, the concrete responsibilities of the owner to the animal be unequivocally stated; (3) proper organizations have standing to represent animals and enforce their legal rights; and (4) the law be regularly evaluated and updated to address the new scientific knowledge in the fields of animal welfare and health.

A brief summary of the five approaches

Each of the five approaches discussed above targets existing deficiencies in the current legal approach to animals and suggests ways to correct them. The approaches differ in terms of the scope of the proposed change – from minimal to fundamental. The approaches also differ in terms of the underlying philosophical view and ultimate goal sought, from Sunstein's approach, falling squarely in the welfarist camp, to Francione's approach, falling strictly within the rights camp.

It is likely that Sunstein's approach is the most politically feasible as it retains the existing paradigm and works to improve the current standards and their enforcement without altering either the fundamental status of animals or the basic requirements of legislation that affects animals. Favre's approach allows for more fundamental alteration of the animals' legal characterization, but it does so under the existing legal paradigm – adding a category of "living property" to the current property scheme. This alteration would not be revolutionary since the law already recognizes a variety of different classes of property, including real, personal, intellectual, and cultural. However, Favre's approach recognizes animals as subjects in a limited sense while allowing for certain human uses, but also provides some protection for their inherent interests. Wise's approach is more controversial than that of Sunstein or Favre. He proposes personhood status for animals, but only for those species who share a high level of practical autonomy with humans (at least initially). Nevertheless, his approach grants equality to all sentient beings and helps to eradicate speciesism from our law by focusing the protection of liberty interests on a non-speciesist and arguably relevant characteristic, practical autonomy. Francione's approach is the most controversial as his ultimate goal is to grant personhood status to all sentient beings and abolish all human use of animals. As a political matter, it is hard to envision a world where no animals are owned or used by humans. However, the criteria he defines for incremental change can be useful in drafting laws to create bright-line rules designed to promote the inherent interests of animals even if some animals remain the property of humans. Finally, GAPA provides a comprehensive set of parameters that offset the speciesist tendencies of the current legal system and address many of the basic problems we noted at the beginning of the chapter.

The future: an all-inclusive legal regime

This final section is designed to draw from our discussion thus far and identify a few of the basic requirements necessary to advance an all-inclusive legal regime where all sentient beings are treated, at least partially, as subjects under the law. Recall our brief discussion of the two philosophical views and associated ultimate goals for how the law should treat animals – welfare and rights. As we have discussed, philosophers, legal scholars, and advocates all interested in protecting animals' interests continue to debate which fundamental view is proper. This section will not resolve that debate as it is beyond the scope of this text. Instead, the following will first identify certain requirements that are consistent with both views, and then present how a legal regime would differ depending upon the view followed.

Some basics to create an all-inclusive legal paradigm

Sentience as a minimum: appreciating sameness and difference

To begin, proponents for animals in both camps generally agree that sentience is the one characteristic necessary for moral consideration. The ability to feel pain and to suffer is why the law should protect *all* sentient beings from the intentional infliction of pain and suffering independent of their relationship to humans. Moreover, scientific proof that an act causes an animal pain and suffering is not required. If animals are deprived of their natural surroundings or of their ability to engage in natural behaviors, the law should presume their welfare is hindered and prohibit such acts. Since pain and suffering is subjective for both humans and nonhumans, Linzey is correct in asserting that arguments such as "we can't really know," "we must have scientific proof," or "they may feel pain, but not as we do," misdirect the debate by minimizing, obfuscating, or belittling animals' suffering.[38]

However, to be truly non-speciesist and all-inclusive, the law must recognize not only the interests that other species share with humans, but also their different and unique interests. Equal consideration very broadly requires that the varied interests of all be counted; that the interests shared by all be weighed equally in calculating the rightness of an action and the boundaries of the law; and in turn, that the law not universally favor human interests over all other species' interests. The ability to understand and respect morally relevant differences while avoiding illegitimate discriminatory treatment has presented a problem for our society and the law throughout time. Whether it be differences based on race, sex, disability, or species, our law continues to struggle with giving true "equal consideration" to all. The traditional "equality" paradigm seeks to treat the minority class equal to the majority class. That is, it sets the interests of the majority as the baseline and protects those same interests of the minority group. However, this equality approach prejudices the interests of the majority over those of the minority. If the minority group has interests or needs different from those of the majority, those interests or needs are often ignored. These problems are of course magnified when addressing differences among species. Nevertheless, a non-speciesist legal regime should recognize the many different species of sentient beings, each with unique capabilities, needs, and desires – for example, the capacity to fly or the need to forage – and appropriately account for them.

Inclusion at every level of the law

Next, animals should be expressly recognized and included within every source and level of the law. (1) Globally, international peremptory rules that now recognize and protect fundamental *human* rights should be expanded to recognize and protect fundamental rights of all sentient beings. Arguably, for example, sentience would justify the basic right to be free from torture and genocide, and thus universal *jus cogens* provisions now reserved for humans only should be extended to include all sentient species. (2) This same principle should be reflected in a society's constitution – the document that is the supreme law of the land and that defines the most fundamental principles on which a society is based. The constitution should expressly recognize that all sentient beings have inherent worth and deserve equal consideration under the law. (3) The criminal cruelty laws should set the bare minimum protection against intentional cruelty and protect all sentient animals with no exceptions. The criminal laws must be robustly enforced with mandatory minimum sanctions sufficient to punish and deter the wrongful conduct. (4) The public statutory and regulatory laws should establish standards that promote the well-being of all animals, independent of their relationship to humans. Sanctions for violations of all public laws should include a minimum level of sanction. (5) Private law should recognize the inherent worth of animals to themselves as well as to their owners when valuing harm, and should compensate the animal as well as the animal's owner. The compensation for the animal's harm to him- or herself would be placed in trust for the benefit of the animal, not their owner. Moreover, private law should take into account the animals' interests when deciding issues governing their fate, such as custody or as beneficiaries of a trust, by granting an animal a guardian ad litem to represent their independent and inherent interests in court.

Representation in every branch

Third, animals must be given a voice within our society. The right to vote is the primary means that humans are given a voice in our legal system. Arguably, animals may have some indirect voice in the "vote" through humans who vote for candidates based in part on their commitment to protect animals' interests. In addition, animals must be given a voice within all three branches of the legal system. First, animals must be granted a voice in the judiciary. Virtually all animal protection legal scholars argue for standing for animals in a court of law. This is essential if the law is to provide meaningful protection to animals. Second, animals' interests must be reflected within the executive branch. This could be accomplished by establishing a separate administrative agency tasked with the *sole* purpose of representing and protecting animals' interests. This agency would be adequately funded and given the same authority and respect as agencies that represent competing human interests. Finally, animals should be granted a voice in the legislature. Currently, the interests of animals are given some voice through humane organizations and individual activists who engage in lobbying efforts to persuade legislators to protect animals' interests. However, as mentioned above, they currently have no direct vote in the selection of representatives.

Legal drafting guidelines

Fourth, when drafting laws to protect animals the following considerations must be taken into account. (1) The animals must be referenced by their species and not characterized by their use or human perception of them. Thus, terms such as "farmed animal," "research animal," "pest," and "nuisance wildlife," would be prohibited. Ethically sensitive terms are necessary to help reorient our view of other species. (2) Terms describing the purpose of the laws must reflect an appreciation for the inherent interests of the animal. Terms such as "wildlife conservation or management" and "animal control" would be replaced with terms such as "wildlife or animal protection." (3) The laws should be drafted as per se bright-line rules, giving no individual discretion to determine the standard defined. Thus, use of terms such as "necessary," "reasonable," or "when practical," would be prohibited.

Utilize the law to create economic incentives, require disclosure and support public education to promote animals' interests

Fifth, as a practical and political matter the law must be used strategically to provide economic incentives, require disclosure, and support educational efforts that promote animals' interests in order to alter societal views and make the enactment and enforcement of an all-inclusive legal regime feasible. Here are a few examples. (1) Provide tax incentives and/or government subsidies to entities who invent, produce and market products that reduce and ultimately replace the use of animals for food or research. Such programs will create the economic incentive and market for the development and sale of compassionate alternatives. (2) Require transparency in the use and treatment of animals. For example, all products derived from animals should contain a label that identifies the animals used and conditions under which they were kept. This will allow consumers to make informed decisions and provide increased incentives for the industry to alter their use of animals based on consumer input. Moreover, industries or government agencies that use or care for animals must be required to publicly disclose statistics on their use and disposal of all animals under their care, including the agricultural industry, research institutions, and animal shelters. Such information will raise the public consciousness and alter societal views of our use and abuse of all animals. (3) Mandate humane education in our public schools to teach compassion and respect for all sentient beings and provide economic incentives for public education campaigns that alert consumers to health concerns, either proven or suspected on the basis of research, for humans from various animal products and that promote veganism.

Welfare versus rights

Finally, as described above, the ultimate goals of the two philosophical views supporting moral consideration for sentient beings are different. A simplified "welfarist" view allows for humans to use animals so long as the animals are treated humanely. A simplified "rights" view seeks also to abolish the use of animals. Thus, how would a welfarist legal regime differ, fundamentally, from a rights legal regime? A welfarist legal regime would retain the current property status of animals while providing greater protections for animals' well-being and providing for greater enforcement of the provision, pursuant to the guidelines discussed. In other words, a welfarist regime would allow the use of animals by humans for human benefit but would impose greater constraints on when, why and how animals may be used. The following guidelines should govern such a regime. (1) The default position should be no use of any sentient being unless specifically allowed by law. (2) The selection criteria to determine which sentient beings may be used should be a function of a non-speciesist characteristic, such as "practical autonomy." (3) The limited uses of only certain limited sentient beings would be allowed, but only when the benefits to all sentient beings significantly outweigh the costs to the sentient beings who are used.

A rights regime would abolish all uses of sentient beings. This would be accomplished by fundamentally altering their legal status as property and granting them the status of "person" under the law. Note that person in this context does not equate to "human." In this context, granting all sentient beings the status of legal person means that the law recognizes them as individuals with inherent interests that may not be jeopardized solely to serve the interests of others, specifically humans. Under a rights regime, the law would criminalize the use of and intentional harm to all sentient beings unless justified to protect one's safety or the safety of another sentient being from direct harm.

Examples

Developing the exact contours of a welfarist or rights legal regime is beyond the scope of this introductory text. An entire book and more would be necessary to properly tackle such an endeavor. The goal here is to begin the conversation. The basic guidelines described above are useful to begin the exploration. However, to provide a bit more context, the following are a couple of specific examples of how the guidelines might be applied in drafting specific laws.

Anti-abuse law

The most fundamental law that is necessary under either regime is the anti-cruelty law. These laws grant the most fundamental right of animals to be free from the intentional infliction of pain and suffering. However, as we have discussed, they are flawed in several respects. First, the cruelty law currently addresses both abuse and neglect but, as we have seen, this often results in exempting wild animals from both. Thus, the law must separately address abuse and neglect: laws against abuse should govern all animals, and neglect laws would govern only domestic animals. Let's discuss how an anti-abuse law would be drafted to be non-speciesist and all-inclusive in order to protect all animals, wild and domestic, from abuse at the hands of humans.

Criminal anti-abuse law

Offense: It is a felony for a human to knowingly or recklessly abuse any nonhuman sentient being, wild or domestic. Abuse includes harming any nonhuman sentient being by torturing, tormenting, starving, chaining, overworking, beating, mutilating, or otherwise causing any sentient being to suffer physical or emotional pain or suffering. It is an aggravated felony for a human to engage in said conduct when acting with malicious intent.

Affirmative defenses: A human will not be held liable for abuse when the abuse was caused in order to protect said human or another sentient being from severe physical injury or when the human was acting to directly benefit the said nonhuman sentient being's individual interests.

Enforcement:

Public: It shall be the responsibility of law enforcement to enforce the anti-abuse law aggressively. The penalty for a felony conviction shall be no less than $X fine and/or X days' imprisonment and no more than $Y fine and/or Y years' imprisonment. The penalty for an aggravated felony conviction shall be no less than $W fine and/or W years' imprisonment and no more than $Z fine and/or Z years' imprisonment. The exact sanction must be determined as appropriate for each society but must be severe enough to reflect the seriousness of the crime and deter and punish the abuser, giving true equal consideration to the interests of the sentient beings who are the victims of such abuse.

Private: Any nonhuman sentient being abused by a human shall have a claim against said human for monetary and injunctive relief. Such relief shall include medical care to treat their injuries and transfer to a safe haven where they will be cared for and provided a good quality of life, including welfare, health and integrity, as defined by species, in detailed, objective specifications. The monetary relief shall compensate for the costs associated with such care and treatment until a permanent placement that provides a good quality of life including welfare, health and integrity, as defined by species, in detailed, objective specifications for the nonhuman sentient being is found. If the sentient being lives in the wild, the ultimate goal when physically and medically feasible, is to treat and care for the sentient being until they may be released to their original home in the wild. The nonhuman sentient being may be euthanized only when in the nonhuman sentient being's individual best interests; for example, when the nonhuman sentient being is injured beyond treatment or rehabilitation.

Registry: Anyone convicted of abuse shall be identified in a public registry.

This anti-abuse law is distinguished from current anti-abuse laws in a number of respects. First, there are no exemptions for certain types of animals or human uses of animals. Thus, all nonhuman sentient beings are protected from individual and institutional instances of abuse. The terms are bright-line, criminalizing all conduct that causes harm to the sentient being unless the affirmative defenses are established. The public sanction includes minimum as well as maximum sanctions designed to punish and deter the wrongful conduct along with public disclosure of conviction through a registry. A private right of action is granted to allow the sentient being whose interests are protected under the law standing to secure enforcement.

Moving from a welfarist to a rights regime

The current law is welfarist and allows humans to use animals for a variety of purposes including breeding and confining huge numbers of animals so that we can then kill them and eat their bodies, none of which are essential to humans' well-being. Nevertheless, use of animals by humans is fundamentally embedded within society and, as a practical and political matter, will take many years to transform and achieve a strict rights legal regime where sentient animals are not used by humans. How might the law facilitate this transformation? Let's explore, in one example, research, the legal elements necessary to achieve the replacement of all live sentient beings used for research.

Research: welfare to rights paradigm

Sunset: No sentient being may be used for research or testing purposes after year Z. Until such time, the government shall promote the development of alternative research and testing techniques that use no sentient beings by providing grants, economic subsidies, and/or tax incentives, to qualified applicants involved in such development.

All government agencies involved in approving the development, production, sale and/or distribution of products shall phase out and eliminate by year Z all requirements of animal testing and/or animal protocols as a requirement for approval.

Allowed use required protocol: Until year Z limited use of specifically defined classes of animals that strictly meet the following criteria are allowed.

A human may own and use a domestic sentient being with a practical autonomy value at 0.5 or below for research purposes only under the following circumstances:

(a) The research will promote scientific knowledge into the medical cure or treatment of life threatening diseases in sentient beings, and

(b) Alternatives that utilize no sentient beings do not exist, and

(c) The use will cause no more than slight pain and suffering, a score of no more than X based on a scale of pain and suffering, and

(d) The human using the sentient being proves by clear and convincing evidence to an independent agency with expertise in the medical area and with the sole goal of protecting the interests of the sentient being who is used, that (a), (b) and (c) are met, and

(e) The human owner provides the sentient being a good quality of life including welfare, health, and integrity as defined by species, in detailed, objective specifications, while under his/her care and control.

Required disclosure: (1) All products that involve the research or testing on sentient beings shall be clearly labeled to inform the consumer of said use. (2) The use, treatment, and disposal of all sentient beings by research project, including the goal of the project and need for the use of sentient beings, shall be disclosed and made publicly available.

Required fees: All entities utilizing sentient beings for research or testing shall pay an annual fee to the Alternative Methods Fund. Money from such fund will be granted to selected entities that develop and market alternative methods that replace the current use of sentient beings for research and testing.

Enforcement:

Public: An independent agency tasked with oversight for the use of sentient beings and with the sole goal of protecting the sentient beings, shall monitor, and review and make public all research uses of sentient beings for compliance with the law. Any use by an entity found not in compliance with the law shall result in the immediate and permanent suspension of any further research by said entity that uses sentient beings.

Private: A sentient being who is used for research in a manner that violates the law shall have a claim against the human owner for monetary and injunctive relief. Such relief shall include medical care to treat his/her injuries and transfer to a safe haven where he/she will be cared for and provided a good quality of life, including welfare, health and integrity as defined by species, in detailed, objective specifications. The monetary relief shall compensate for the costs associated with such care until a permanent placement that provides a good quality of life, including welfare, health and integrity as defined by species, in detailed, objective specifications, for the sentient being is found. The sentient being may be euthanized only when in the sentient being's own best interests; for example, when the sentient being is injured beyond treatment or rehabilitation.

This example demonstrates application of the 3Rs, explicit requirements approved by an independent agency tasked with protecting the animals' interests, criteria based on objective qualities not species classification, no exemptions, public disclosure, and public and private enforcement schemes that severely sanction any noncompliance with the law. It is true that such use arguably would violate the anti-abuse law. However, note that the law limits use to *domestic* sentient beings with a *practical autonomy* of 0.5 or below. Thus, the law does not differentiate on species but rather on a non-speciesist characteristic. The only use allowed serves a research goal of paramount importance to all sentient beings, the cure or treatment of only life-threatening diseases in all sentient beings. Moreover, the sentient being used may suffer only slightly as determined based on an objective scale of pain and suffering. Further, the rules are bright-line rather than rules of reason. Arguably, some of the terms used are subject to interpretation. However, none of the terms dictate independent discretion. For example, "a good quality of life, including welfare, health and integrity" will be defined by species in detailed specifications that set standards for every need and want, physical, psychological and emotional, of that species. One can still argue that opinions may differ as to when a sentient being is beyond rehabilitation, but that is not a function of the terms. For example, the proposal does not say beyond "reasonable" treatment or rehabilitation, where the term "reasonable" imposes a level of independent discretion. Moreover, there is a definite end to all use of sentient beings for such purpose and the law provides both a carrot and stick to incentivize replacing the use of sentient beings in the interim years by providing economic incentives to create compassionate alternatives that do not use sentient beings and imposing a fee on the continued use of sentient beings for such use. Admittedly, the law is only effective if enforced and the law provides for both public and private enforcement. Public enforcement by an agency with the sole goal of protecting sentient beings (not promoting research) and a draconian sanction for any violation – the suspension of all such research. The private enforcement grants the sentient being standing to enforce the law and full compensation for the harm inflicted.

Conclusion

Animals are the property of humans under the law, and humans own and use animals for a variety of purposes. Individual instances of gratuitous intentional cruelty against certain animals are banned, while institutionalized abuse of animals is allowed and often promoted under the law. The current legal regime, drafted by and for humans, is speciesist, organized around such uses, and is designed to provide minimal standards of care for certain animals so long as the intended human use of the animal is not jeopardized. As such, human interests virtually always trump the interests of animals.

Over time, societal views of animals and their relationship to humans change. Philosophers have debated the proper status of animals throughout history and a growing consensus exists to fundamentally alter the status of animals. The law in turn changes to accommodate these moral and ethical views. Today, the notion of animal law, an area of study and practice once unknown, is now beginning to flourish. Countries are recognizing that animals are sentient beings and are including the protection of animals' inherent interests within their constitutions; states are enacting tougher criminal laws and incentivizing their enforcement; and courts and legislatures are beginning to recognize the true value of companion animals to their owners. Through advocacy, education, and economic incentives, society and the law will move towards a greater appreciation of the animals with whom we share this world, and a non-speciesist, all-inclusive legal regime may develop that gives equal consideration to and protection of the individual, inherent interests of all sentient beings.

Notes

Chapter 1

1. A. Hall, 2008, *European Court Agrees to Hear Chimps Plea for Human Rights* [online], Mail Online. Available from: http://www.dailymail.co.uk/news/worldnews/article-1020986/European-Court-agrees-hear-chimps-plea-human-rights.html [accessed: 10.5.2010].
2. *United States of America* v. *Approximately 53 Pit Bull Dogs*, 3:07CV397 (E.D. Va. Nov. 15, 2007).
3. *Ibid.*
4. F.W.R. Brambell, 1965, *Report of the Technical Committee to Enquire into the Welfare of Animals Kept Under Intensive Livestock Husbandry Systems*, Her Majesty's Stationery Office, London.
5. See generally Department for Environment, Food and Rural Affairs (DEFRA) [online]. Available from: http://www.defra.gov.uk/foodfarm/farmanimal/welfare/onfarm/index.htm [accessed: 10.5.2010].
6. *World Declaration on Great Apes*
   We demand the extension of the community of equals to include all great apes: human beings, chimpanzees, bonobos, gorillas and orangutans. The community of equals is the moral community within which we accept certain basic moral principles or rights as governing our relations with each other and enforceable at law. Among these principles or rights are the following:

   1. The Right to Life
   The lives of members of the community of equals are to be protected. Members of the community of equals may not be killed except in very strictly defined circumstances, for example, self-defense.
   2. The Protection of Individual Liberty
   Members of the community of equals are not to be arbitrarily deprived of their liberty; if they should be imprisoned without due legal process, they have the right to immediate release. The detention of those who have not been convicted of any crime, or of those who are not criminally liable, should be allowed only where it can be shown to be for their own good, or necessary to protect the public from a member of the community who would clearly be a danger to others if at liberty. In such cases, members of the community of equals must have the right to appeal, either directly or, if they lack the relevant capacity, through an advocate, to a judicial tribunal.
   3. The Prohibition of Torture
   The deliberate infliction of severe pain on a member of the community of equals, either wantonly or for an alleged benefit to others, is regarded as torture, and is wrong.

World Declaration on Great Apes [online], Great Ape Protection. Available from: http://www.greatapeproject.org/en-US/oprojetogap/Declaracao/declaracao-mundial-dos-grandes-primatas [accessed: 10.5.2010].

7. As of January 2009 the Spanish Government had not pronounced officially. Thus, the Deputies "sent a new document to the Government asking if the Government would fulfil the proposal and support the Great Ape Project, as it had been required by the Deputies Chamber. The Government apparently never responded and the law was never enacted. Unfortunately no reference could be located explaining why the Government refused to comply with the request from the Deputies although clearly such a law would have been quite controversial as no other country to date has adopted the GAP under their laws. For example, one bioethicist who apparently did not support the law stated: "Of course the purpose of this isn't to merely improve the treatment of great apes – which could be accomplished as it already has been in some places via normal animal welfare statutes. Rather, the explicit point of the GAP is revolutionary – to demote human beings from the uniquely valuable species and into merely another animal in the forest." J. Jalsevac, 2008, *Human Rights for Apes: Spanish Parliament Passes Unprecedented Resolution*, LifeSite [online] (quoting Wesley Smith). Available from: Newshttp://www.lifesitenews.com/ldn/2008/jun/08062605.html.

8. L. Abend, 2008, *In Spain, Human Rights for Apes* [online], Time. Available from: http://www.time.com/time/world/article/0,8599,1824206,00.html [accessed: 10.5.2010].

9. T. Catan, 2008, *Apes get Legal Rights in Spain, to Surprise of Bullfight Critics* [online], Times Online. Available from: http://www.timesonline.co.uk/tol/news/world/europe/article4220884.ece [accessed: 10.5.2010].

10. 2009, *Parliament of Catalonia Bans Bull Fighting* [online], Radio Netherlands Worldwide. Available from: http://www.rnw.nl/english/article/parliament-catalonia-bans-bull-fighting [accessed: 10.5.2010].

11. 2009, *Catalonia to Vote on Bull Fight Ban* [online], Aljazeera. Available from: http://english.aljazeera.net/news/europe/2009/12/2009121817577600314.html [accessed: 10.5.2010].

12. P. Wedderburn, 2009, *China Unveils First Ever Animal Cruelty Legislation* [online], Telegraph. Available from: http://blogs.telegraph.co.uk/news/peter-wedderburn/100010449/china-unveils-first-ever-animal-cruelty-legislation/ [accessed: 10.5.2010].

13. 2004, *Animal Welfare Legislation Debuts in China* [online], Consulate General of the People's Republic of China in Los Angeles. Available from: http://losangeles.china-consulate.org/eng/topnews/t127829.htm [accessed: 14.5.2010].

14. 2009, *China Animal Welfare Law Under Review while 36,000 Dogs Culled* [online], IFAW. Available from: http://www.ifaw.org/ifaw_southern_africa/media_center/press_releases/6_16_2009_55219.php [accessed: 10.5.2010].

15. 2010, *Proposed Animal Welfare Law Watered Down* [online], China.org.cn. Available from: http://www.china.org.cn/china/2010-01/26/content_19309286.htm [accessed: 10.5.2010]

16. A. Linzey and P. Waldau, 1995, "Speciesism," in *Dictionary of Ethics, Theology and Society*, at 788, Her Majesty's Stationery Office, Routledge.

17. J. Dunayer, 2004, *Speciesism*, at 2, Lantern Books, Routledge (quoting P. Singer, 1975, *Animal Liberation*, at 30, HarperCollins, New York).
18. *Ibid.* at 3.
19. A. Linzey, 2009, *Why Animal Suffering Matters*, Oxford University Press, New York.
20. *Ibid.* at 52.
21. *Ibid.* (quoting *Codes of Recommendations for the Welfare of Livestock: Report by the Farm Animal Welfare Advisory Committee* 5, London, MAFF, 1970).
22. S. Waisman, P. Frasch and B. Wagman, 2006, *Animal Law: Cases and Materials*, 3rd edn, at ch. 27, Carolina Academic Press, North Carolina.
23. The United States is made up of 50 states and the District of Columbia.
24. The United Kingdom refers to England, Scotland, Wales and Northern Ireland.
25. Australasia refers to Australia, New Zealand, the island of New Guinea, and neighboring islands in the Pacific Ocean. The Commonwealth of Australia is made up of six states – Western Australia, Northern Australia, South Australia, Queensland, New South Wales, and Victoria – and two major mainland territories.
26. The European Union refers to the 27 member states of Austria, Belgium, Bulgaria, Cyprus, Czech Republic, Denmark, Estonia, Finland, France, Germany, Greece, Hungary, Ireland, Italy, Latvia, Lithuania, Luxembourg, Malta, Netherlands, Poland, Portugal, Romania, Slovakia, Slovenia, Spain, Sweden, and the United Kingdom.
27. G. Francione, 2008, *Animals as Persons*, at 138, Columbia University Press, New York (quoting C. Darwin, 1981, *The Descent of Man,* at 105, Princeton University Press, New Jersey).
28. M. Bekoff and J. Pierce, 2009, *Wild Justice: The Morals of Lives of Animals*, at x, University of Chicago Press, Chicago.
29. *Ibid.* at xi.
30. See for example *Can Animals Predict Disaster?* [online], Nature. Available from: http://www.pbs.org/wnet/nature/episodes/can-animals-predict-disaster/introduction/130/ (recounting the wild trumpeting of an elephant chained to a tree who breaks free to flee to higher ground only moments before a massive tsunami crashes the shore) [accessed at 13.5.2010].
31. Linzey, *Why Animal Suffering Matters*, at 17.
32. *Ibid.* at 19.
33. *Ibid.* at 20.
34. *Ibid.* at 21–22.
35. S. Otto, 2010, *Animal Protection Laws of the USA and Canada* [online], Animal Legal Defense Fund. Available from: http://www.aldf.org/article.php?id=259 [accessed: 14.5.2010].
36. Europa (1997) Treaty of Amsterdam [online], Amsterdam. Available from: http://eur-lex.europa.eu/en/treaties/dat/11997D/htm/11997D.html [accessed: 10.5.2010]; Europa (2007) Treaty of Lisbon [online], Lisbon. Available from: http://europa.eu/lisbon_treaty/full_text/index_en.htm [accessed 14.5.2010].
37. Aristotle, 1999, *Politics*, at 13 [online], trans. B. Jowett, Kitchener, Batoche Books. Available from: http://socserv.socsci.mcmaster.ca/oldecon/ugcm/3ll3/aristotle/Politics.pdf.

38. I. Kant, 1963 [1775–1780], *Lectures on Ethics*, 240, Hackett Publishing, Cambridge (based on manuscripts from three of Kant's students).
39. Some commentators consider the term "wild" to be offensive, connoting a being as violent, and thus prefer to reference undomesticated animals as "free-ranging." This text, although sensitive to language used to describe animals as discussed above, will refer to such animals as "wild" because the term is so deeply entrenched in the law. Moreover, as a legal matter, "free-ranging" may refer to a domesticated animal that is not confined, for example a free-ranging chicken or cat. Thus, to alter this term would create serious problems of interpretation. Finally, this author does not find the term "wild" as offensive as other terms used to describe animals, such as "brute" or "pest," and thus believes that the need for such a change of terminology in this text is not warranted.
40. A. Cassese, 2001, *International Law*, at 4–5, Oxford University Press, New York.
41. *Ibid.* at 141.
42. In a world where animals are recognized as sentient beings with inherent worth, their fundamental rights might be included as well. In fact, the GAP is formulated to apply *jus cogens* to the great apes.
43. Council of Europe, Convention on Civil Liability for Damage Resulting from Activities Dangerous to the Environment [online]. Available from: http://conventions.coe.int/treaty/en/treaties/html/150.htm; see generally Cassese, *International Law*, ch. 17, Protection of the Environment.
44. Cassese, *International Law*, at 173.
45. See below, notes 70–73.
46. See generally Peter E. Quint, "What is a Twentieth-Century Constitution?," 67 *Md. L. Rev.* 238 (2007).
47. Claudia E. Haupt, "The Nature and Effects of Constitutional State Objectives: Assessing the German Basic Law's Animal Protection Clause," 16 *Animal L.* 213, 222 (2010).
48. Bruce Cain and Roger Noll, "Malleable Constitutions: Reflections on State Constitutional Reform," 87 *Tex. L. Rev.* 1517, 1518 (2009).
49. *Ibid.*
50. State constitutions can be changed in two ways: amended or revised. Generally, a fundamental change to the constitution is deemed a revision and is harder to enact. However, there is no bright-line demarking the two. *Ibid.* at 1524.
51. D. Polhill, "Democracy's Journey," in M. Dane Waters, ed., 2001, *The Battle Over Citizen Lawmaking*, at 10, Carolina Academic Press, Durham.
52. Generally, 6–10 percent of the prior gubernatorial vote. Cain & Noll, "Malleable Constitutions," at 1523.
53. *Ibid.* at 1530.
54. W. Pacelle, "The Animal Protection Movement: A Modern-Day Model Use of the Initiative Process," in Waters, *The Battle Over Citizen Lawmaking*, at 113.
55. The Netherlands and France were unwilling to adopt a constitutional treaty without major amendments. Moreover, the United Kingdom pressed for an "amending treaty" that could be ratified by means of a parliamentary vote. Thus the European Union has no formal constitution. See *Q&A: The*

Lisbon Treaty [online], BBC News. Available from: http://news.bbc.co.uk/2/ hi/europe/6901353.stm [accessed: 11.24.2009]. A few countries negotiated the ability to "opt out" of specific provisions of the final treaty as well. *Ibid.*
56. Europa (2007) Treaty of Lisbon [online], Lisbon. Available from: http:// eur-lex.europa.eu/JOHtml.do?uri=OJ:C:2007:306:SOM:EN:HTML.
57. D. Favre and V. Tsang, 1993, *The Development of the Anti-Cruelty Laws During the 1800s*, at 1 [online], Animal Legal & Historical Center, Available from: http://www.animallaw.info/articles/arusfavrehistcruelty1993.htm#Early%20 American%20Legislation [accessed: 12.5.2010].
58. G. Becker, 2009, *The Animal Welfare Act: Background and Selected Legislation*, at 103 [online], Congressional Research Service. Available from: http://www. nationalaglawcenter.org/assets/crs/RS22493.pdf [accessed: 12.5.2010].
59. See generally 2008 C.F.R. Title 9. Available from: http://www.access.gpo.gov/ nara/cfr/waisidx_08/9cfr3_08.html.
60. 7 U.S.C. §§ 1901–1907.
61. 16 U.S.C. §§ 1531–1534.
62. See generally DEFRA [online]. Available from: http://ww2.defra.gov.uk/.
63. *Environment* [online], DEFRA. Available from: http://www.defra.gov.uk/ environment/index.htm.
64. *Food and Farming* [online], DEFRA. Available from: http://www.defra.gov.uk/ foodfarm/index.htm.
65. *Wildlife and Pets* [online], DEFRA. Available from: http://www.defra.gov.uk/ wildlife-pets/index.htm.
66. See for example *On-Farm Animal Welfare* [online], DEFRA. Available from: http://www.defra.gov.uk/foodfarm/farmanimal/welfare/onfarm/index. htm#we.
67. See for example *Companion Animal Welfare Council* [online]. Available from: http://www.cawc.org.uk/.
68. See *Research and Testing Using Animals* [online], U.K. Home Office. Available from: http://www.homeoffice.gov.uk/science-research/animal-research/.
69. See generally *Understanding Animal Research* [online]. Available from: http:// www.understandinganimalresearch.org.uk/homepage.
70. *Decision-making in the European Union* [online], Europa. Available from: http:// europa.eu/institutions/decision-making/index_en.htm [accessed: 12.5.2010].
71. 2010, *Consolidated Version of the Treaty on the Functioning of the European Union*, art. 288 [online], Europa, Available from: http://eur-lex.europa.eu/ LexUriServ/LexUriServ.do?uri=OJ:C:2008:115:0047:0199:EN:PDF [accessed: 12.5.2010].
72. 2010, *Process and Players*, 1.3.1 [online], Europa, Available from: http://eur-lex. europa.eu/en/droit_communautaire/droit_communautaire.htm#1.3.31 [accessed: 12.5.2010].
73. See generally *Animal Health and Welfare* [online], Europa. Available from: http://ec.europa.eu/food/animal/welfare/references_en.htm.
74. R. Wacks, 2006, *Philosophy of Law: A Very Short Introduction*, at 53–55, Oxford University Press, New York.
75. *Ibid.*
76. *Geer* v. *Connecticut*, 161 U.S. 519, 524 (1896).
77. *Ibid.* at 526–527 (quoting W. Blackstone, 2001, *Blackstone's Commentaries on the Laws of England*, Cavendish Publishing, Great Britain).

78. *Ibid.* at 526 (quoting Napoleonic Code arts. 714, 715).
79. *Ibid.* at 529.
80. *Moore* v. *Regents Univ. of California*, 793 P.2d 479, 509 (Cal. 1990).
81. *Ibid.*

Chapter 2

1. R.C.W. § 16.52.205(2).
2. A. Linzey, 2009, *Why Animal Suffering Matters*, 9–10, Oxford Press, London.
3. F.S.A § 828.13.
4. *Ibid.* § 828.12.
5. 510 I.L.C.S. § 70/2.0.
6. *Ibid.* § 2.06. "Owner" means any person who (a) has a right of property in an animal, (b) keeps or harbors an animal, (c) has an animal in his care, or (d) acts as custodian of an animal.
7. *Ibid.* § 3.
8. *Ibid.* § 2.07. "Person means any individual, minor, firm, corporation, partnership, other business unit, society, association, or other legal entity, any public or private institution, the State of Illinois, or any municipal corporation or political subdivision of the State."
9. *Ibid.* § 3.01.
10. *Ibid.* § 2.01a. "Companion Animal" means an animal that is commonly considered to be, or is considered by the owner to be, a pet. "Companion Animal" includes, but is not limited to, canines, felines, and equines.
11. *Ibid.* § 3.02.
12. *Ibid.* § 3.03.
13. *Ibid.* § 2.10.
14. See generally *The Hoarding of Animals Research Consortium* [online]. Available from: http://www.tufts.edu/vet/cfa/hoarding/ [accessed: 5.25.2010]. The site states: "In a typical hoarding situation, the hoarder will put their own needs to be surrounded by animals ahead of providing even the most basic care. Although professing great love for the animals, they are often oblivious to serious illness, animals in desperate need of veterinary care, starvation, and even death of the animals. Few if any animals are ever adopted or placed, and new animals are never turned away, even in the face of rapidly deteriorating conditions. There are often substantial efforts to acquire even more pets."
15. *Ibid.* ("Hoarding is characterized in DSM-IV (the bible of psychiatric disorders) as a symptom of obsessive-compulsive disorder (OCD) and obsessive-compulsive personality disorder (OCPD).")
16. United Kingdom Animal Welfare Act 2006 (ch. 45, § 9(2)), London, DEFRA.
17. New Zealand Animal Welfare Act 1999 § 29(a), Wellington, MAF.
18. Animal Welfare Amendment Bill 2010 (118–1). Available from: http://www.legislation.govt.nz/bill/government/2010/0118/latest/whole.html; see 2010, *The Government has Introduced Legislation Increasing the Penalties for Animal Cruelty* [online], Radio New Zealand. Available from: http://www.radionz.co.nz/news/stories/2010/02/16/1247f23cfc0c [accessed: 5.25.2010].
19. Animal Welfare Amendment Bill 2010 (118–1), Explanatory Note. Available from: http://www.legislation.govt.nz/bill/government/2010/0118/9.0/DLM2747701.html.

20. P. Sankoff, 2009, "The Welfare Paradigm," in P. Sankoff and S. White, eds., *Animal Law in Australasia: A New Dialogue*, at 23, Federation Press, Sydney.
21. See *ibid.* at 22–25.
22. M.L. Randour and H. Davidson, 2008, *A Common Bond: Maltreated Children and Animals in the Home*, American Humane Association, Englewood, 7; Randall Lockwood, 1999, "Animal Cruelty and Violence Against Humans: Making the Connection," 5 *Animal L.* 81, 81.
23. P. Arkow, 2003, *Breaking the Cycles of Violence: A Guide to Multi-Disciplinary Interventions*, The Latham Foundation For the Promotion of Humane Education, Alameda, at 17; L.M. Renner and K.S. Stark, 2004, *Intimate Partner Violence and Child Maltreatment: Understanding Co-occurrence and Intergenerational Connections*, Institute for Research on Poverty at University of Wisconsin; E. DeViney et al., 1983, "The Care of Pets Within Abusing Families," 4 *Int'l J. Study of Animal Problems* 321–329; see also F.R. Ascione, 1997, "The Abuse of Animals and Domestic Violence: A National Survey of Shelters for Women who are Battered," *Society and Animals* 5, 205–218.
24. See Judge Roger Dutson, 1994, "Domestic Violence," 7 *Utah B.J.* 42, 43.
25. See generally F. Ascione and P. Arkow, eds., 1999, *Child Abuse, Domestic Violence and Animal Abuse,* Purdue University Press, West Lafayette.
26. L.M. Broidy et al., 2003, "Developmental Trajectories of Childhood Disruptive Behaviors and Adolescent Delinquency: A Six-Site Cross-National Study," 39 *Developmental Psychopathology* 2 (March), 222–245; F. Ascione, 1998, "Battered Women's Reports of Their Partners' and Their Children's Cruelty to Animals," in R. Lockwood and F. Ascione, eds., *Cruelty to Animals and Interpersonal Violence: Readings in Research and Application*, Purdue Research Foundation, Purdue, IL, 293–294.
27. See C.S. Widom, 1989, "The Cycle of Violence," 244 *Science* 160–166; K. Heide and L. Merz-Perez, 2003, *Animal Cruelty: Pathway to Violence Against People*, Altamira Press, Oxford ("Indicating that animal abuse during childhood serves as a "red flag" early on, the study is the first to provide both quantitative and qualitative analyses of the correlation between childhood animal cruelty and adult violent behavior."); see also R. Lockwood and G. Hodge, 1998, "The Tangled Web of Animal Abuse: The Links between Cruelty to Animals and Human Violence," in Lockwood and Ascione, *Cruelty to Animals and Interpersonal Violence*, 77–82; see generally F. Ascione, 2001, *Animal Abuse and Youth Violence* [online], U.S. Department of Justice. Available from: http://www.ncjrs.gov/pdffiles1/ojjdp/188677.pdf [accessed: 5.25.2010].
28. B.J. Walton-Moss, J. Manganelo, V. Frye and J. Campbell, 2005, "Risk Factors for Intimate Partner Violence and Associated Injury among Urban Women," *J. Community Health*, 377–389 (a study done over a seven-year period using control groups and randomization found animal cruelty to be one of four factors that predict who would become a batterer). Also, the Pittsburgh Study, by Rolf Loeber, is an ongoing longitudinal study of the causes and correlates of youth violence that has found cruelty toward people and animals (combined factor) related to the persistence in deviant and aggressive behavior. See Randour and Davidson, *A Common Bond*, at 6.

29. See for example J. Levin and A. Arluke, 2009, "Reducing the Link's False Positive Problem," at 63 in A. Linzey ed., *The Link Between Animal Abuse and Human Violence*, Sussex Academic Press, Eastbourne.
30. C. Siebert, 2010, "The Animal Cruelty Syndrome," *New York Times*, June 7. Available from: http://mail.law.gwu.edu/wm/eml/login.html?sessionid=18 6cfac183c6743c7a4da933fa718dfd4&op=entry.
31. Heide and Merz-Perez, *Animal Cruelty*, at 15.
32. *Ibid.* at 16.
33. A.J. Fitzgerald, L. Kalof and T. Dietz, 2009, "Slaughterhouses and Increased Crime Rates: An Empirical Analysis of the Spillover From 'The Jungle' into the Surrounding Community", *Organization & Environment*, [online] 158–184. Available from: http://oae.sagepub.com/cgi/content/abstract/22/2/158 [accessed: 5.25.2010].
34. *Ibid.* at 163.
35. *Ibid.* at 173.
36. *Hannon* v. *State*, 941 So.2d 1109, 1142–1143 (Fla. 2006).
37. *People* v. *Griffin*, 93 P.3d 344, 373–375 (Cal. 2004).
38. *See generally* J. Schaffner, 2009, "Law and Policy to Address the Link of Family Violence," in A. Linzey ed., *The Link Between Animal Abuse and Human Violence*, Sussex Academic Press, Eastbourne.
39. Puerto Rico is a Commonwealth of the United States, an unincorporated, organized territory.
40. P.R. Animal Protection and Welfare Act (S.B. 2552) [online]. Available from: http://www.oslpr.org/download/en/2008/A-0154-2008.pdf [accessed: 5.25.2010] .
41. "Animal" is defined in section 2(b) as "any mammal, bird, reptile. Amphibian, fish, cetacean, and any other superior phyla animal in captivity or under the control of any person, or any animal protected by Federal or Commonwealth laws or by municipal ordinances." *Ibid.*
42. *Ibid.* § 19. Section 19(b) expressly bans "experiments for education purposes" in grade through high school levels.
43. *People* v. *O'Rourke*, 369 N.Y.S. 2d 335 (Crim. Ct. City N.Y. 1975).
44. The cruelty law stated: "A person who overdrives, overloads, tortures or cruelly beats or unjustifiably injures, maims, mutilates or kills any animal, whether wild or tame, and whether belonging to himself or to another, or deprives any animal of necessary sustenance, food or drink, or neglects or refuses to furnish it such sustenance or drink, or causes, procures or permits any animal to be overdriven, overloaded, tortured, cruelly beaten, or unjustifiably injured, maimed, mutilated or killed, or to be deprived of necessary food or drink, or who willfully sets on foot, instigates, engages in, or in any way furthers any act or cruelty to any animal, or any act tending to produce such cruelty, is guilty of a misdemeanor, punishable by imprisonment for not more than one year, or by a fine of not more than five hundred dollars, or by both." N.Y. Agr. & Markets Law § 353.
45. *Ibid.*
46. *People* v. *Walsh*, 19 Misc.3d 1105(A) *3–4 (NYC Crim. Ct. 2008) (unreported decision).
47. 369 N.Y.S.2d at 341.
48. *Ibid.* at 341–342.

49. See for example *Cock Fighting Fact Sheet* [online], HSUS. Available from: http://www.hsus.org/hsus_field/animal_fighting_the_final_round/cockfighting_fact_sheet/ [accessed: 5.25.2010].

50. See for example *Dog Fighting Fact Sheet* [online], HSUS. Available from: http://www.hsus.org/hsus_field/animal_fighting_the_final_round/dogfighting_fact_sheet/ [accessed: 5.25.2010]; *Dog Fighting FAQ* [online], ASPCA. Available from: http://www.aspca.org/fight-animal-cruelty/dog-fighting/dog-fighting-faq.html [accessed: 5.25.2010] .

51. Va. Code Ann. § 3.2-6571:

> A. No person shall knowingly:
> 1. Promote, prepare for, engage in, or be employed in, the fighting of animals for amusement, sport or gain;
> 2. Attend an exhibition of the fighting of animals;
> 3. Authorize or allow any person to undertake any act described in this section on any premises under his charge or control; or
> 4. Aid or abet any such acts.
> Except as provided in subsection B, any person who violates any provision of this subsection is guilty of a Class 1 misdemeanor.
> B. Any person who violates any provision of subsection A in combination with one or more of the following is guilty of a Class 6 felony:
> 1. When a dog is one of the animals;
> 2. When any device or substance intended to enhance an animal's ability to fight or to inflict injury upon another animal is used, or possessed with intent to use it for such purpose;
> 3. When money or anything of value is wagered on the result of such fighting;
> 4. When money or anything of value is paid or received for the admission of a person to a place for animal fighting;
> 5. When any animal is possessed, owned, trained, transported, or sold with the intent that the animal engage in an exhibition of fighting with another animal; or
> 6. When he permits or causes a minor to (i) attend an exhibition of the fighting of any animals or (ii) undertake or be involved in any act described in this subsection.

52. See for example *Fact Sheet, Dogfighting: State Laws* [online], HSUS. Available from: http://www.hsus.org/acf/fighting/dogfight/ranking_state_dogfighting_laws.html [accessed: 5.25.2010]; *Fact Sheet, Cockfighting: State Laws* [online], HSUS. Available from: http://www.hsus.org/web-files/PDF/animal_fighting/cockfighting_statelaws.pdf [accessed: 5.25.2010].

53. Hog-dog fighting is a less common form of animal fighting in the U.S. "In a hog-dog fight or 'hog-dog rodeo,' a trained dog attacks a trapped feral hog who is released into an enclosed pit from which there is no escape. To confer an advantage on the dog, fight organizers will either cut off the hog's tusks or outfit the dog in a Kevlar vest. While the crowd cheers, the dog is timed to see how quickly he can pin down the hog by tearing into the hog's snout, ears and eyes." Available at: http://animalabusersspotlight.yolasite.com/dog-and-cock-fighting.php.

54. See generally Addie Patricia Asay, 2003, "Greyhounds: Racing to Their Death," 32 *Stet. L. Rev.* 433, *Grey2K USA* [online]. Available from: http://www.grey2kusa.org/racing/cruel.html [accessed: 5.25.2010].
55. 7 U.S.C. § 2156 (a)(b). "The term 'animal fighting venture' means any event, in or affecting interstate or foreign commerce, that involves a fight conducted or to be conducted between at least 2 animals for purposes of sport, wagering, or entertainment, except that the term 'animal fighting venture' shall not be deemed to include any activity the primary purpose of which involves the use of one or more animals in hunting another animal"; and "the term 'instrumentality of interstate commerce' means any written, wire, radio, television or other form of communication in, or using a facility of, interstate commerce." *Ibid.* (g)(1)(2).
56. Animal Welfare Act, ch. 45, § 8. The first attempt to outlaw animal fighting in the United Kingdom was in 1800.
57. *Ibid.* § 8(7).
58. *Ibid.* § 8(3).
59. The Animal Welfare Act states in relevant part that:

   (c) ... It shall be unlawful for any person to knowingly use the mail service of the United States Postal Service or any instrumentality of interstate commerce for commercial speech for purposes of advertising an animal, or an instrument described in subsection (e), for use in an animal fighting venture, promoting or in any other manner furthering an animal fighting venture except as performed outside the limits of the States of the United States.
   (d) ... Notwithstanding the provisions of subsection (c) of this section, the activities prohibited by such subsection shall be unlawful with respect to fighting ventures involving live birds only if the fight is to take place in a State where it would be in violation of the laws thereof.
   (e) ... It shall be unlawful for any person to knowingly sell, buy, transport, or deliver in interstate or foreign commerce a knife, a gaff, or any other sharp instrument attached, or designed or intended to be attached, to the leg of a bird for use in an animal fighting venture. 7 U.S.C. § 2156 (as amended on June 18, 2008)

   The Postal Reorganization Act makes "[m]atter the deposit of which in the mails is punishable under ... section 26 of the Animal Welfare Act nonmailable." 39 U.S.C. § 3001.
60. 18 U.S.C. § 48 (a)(b).
61. See H.R. Rep No. 106–397 at 2 (1999).
62. *United States* v. *Stevens*, 533 F.3d 218, 221 (3rd Cir. 2008).
63. *United States* v. *Stevens*, 130 S. Ct. 1577 (2010).
64. *Ibid.* at 1587.
65. *Stevens*, 533 F.3d at 233.
66. *Ibid.* at 236 (Cowen, J. dissenting). Conduct may also be expressive and then subject to some protection under the First Amendment. In the United States, nude dance clubs, for example, have been the grist for several cases to address whether local jurisdictions may outlaw these "expressive" entertainment establishments. Judge Posner, in *Miller* v. *City of South Bend*, 904 F.2d 1081, 1097 (7th Cir. 1990), in his concurrence suggested an interesting comparison to the issue, bull-fighting. He explained that bull-fighting "is

an expressive activity and even has affinities to the dance ... orchestrated and choreographed for maximum emotional impact; among the feelings conveyed are grace, courage, suffering, fear, beauty, cruelty, splendor, and *machismo*. Hemingway, *Death in the Afternoon* (1932)." Nevertheless, it is not protected by the First Amendment because the "Constitution does not place freedom of expression above all other values." *Ibid*. There are many grounds for prohibiting bull-fighting, including "aversion to mutilating or killing animals for sport," which in this case outweighs its expressive value. *Ibid*.

67. *Stevens*, 130 S. Ct. at 1588.
68. *Ibid*. at 1590.
69. *Ibid*. at 1592.
70. *Ibid*. (Alito, J. dissenting).
71. *Ibid*. at 1595 (Alito, J. dissenting).
72. *Ibid*. at 1600 (Alito, J. dissenting: emphasis added).
73. *Ibid*. at 1601 (Alito, J. dissenting: emphasis added).
74. L. McCormick, 2010, *Bill to Ban Animal Crush Videos Clears First Hurdle*, ConsumerAffairs.com, [online]. Available from: http://www.consumeraffairs. com/news04/2010/06/animal_cruelty_bill.html [accessed: 9.16.2010].
75. H.R. 5566, 11th Cong. 2d Sess. (2010). The Judiciary Committee of the House approved the bill in late June 2010.
76. *Ibid*. (emphasis added).
77. 18 U.S.C. § 43(a) (emphasis added).
78. K. McCoy, 2007, "Subverting Justice: An Indictment of the Animal Enterprise Terrorism Act," 14 *Animal L.* 53.
79. See for example R. Brisbin and S. Hunter, 2008, *Religion, Animal Rights and Animal Legislation* [online]. Available from: http://www.polsci.wvu.edu/ faculty/BRISBIN/Papers/2008.%20Religion%20&%20Animals.pdf [accessed: 5.25.2010].
80. *Church of the Lukumi Babalu Aye* v. *City of Hialeah*, 508 U.S. 520, 532 (1993).
81. See generally T. Bryant, D. Cassuto and R. Huss, eds., 2008, *Animal Law and the Courts: A Reader* at 36–77, Thompson West, Eagan.
82. *Lukumi Babalu*, 508 U.S. at 524–525.
83. *Ibid*. at 527.
84. *Ibid*. at 528.
85. 494 U.S. 872 (1990).
86. *Lukumi Babalu*, 508 U.S. at 531.
87. *Ibid*. at 539.
88. *Ibid*. at 580 (Blackmun, J. concurring).
89. *Ibid*. at 559 (Scalia, J. concurring in part).
90. Cha'are Shalom Ve Tsedek, 2000-VII Eur. Ct. H.R. 232.
91. See generally Claudia E. Haupt, 2007, "Free Exercise of Religion and Animal Protection: A Comparative Perspective on Ritual Slaughter," 39 *Geo. Wash. Int'l L. Rev.* 839, 849–856.
92. Convention for the Protection of Human Rights and Fundamental Freedoms art. 9, Sept. 3, 1953, 213 U.N.T.S. 222.
    Article 9. Freedom of thought, conscience and religion
    (1) Everyone has the right to freedom of thought, conscience and religion; this right includes freedom to change his religion or belief and freedom, either alone or in community with others and in public or private,

to manifest his religion or belief, in worship, teaching, practice and observance.

(2) Freedom to manifest one's religion or beliefs shall be subject only to such limitations as are prescribed by law and are necessary in a democratic society in the interests of public safety, for the protection of public order, health or morals, or for the protection of the rights and freedoms of others.

93. 374 F. Supp. 1284 (S.D.N.Y. 1974). This case is recognized as "the first animal rights lawsuit" to be brought in the United States. Joyce Tischler, 2008, "The History of Animal Law Part I, (1972–1987)," 1 *Stanford J. Animal L. & Policy* 4. Available from: http://www.aldf.org/downloads/Tischler_StanfordJournal-Vol1.pdf. Note that although the HMSA is an animal welfare law and not a criminal law, the competing arguments are similar.

94. 374 F. Supp. 1289.

95. *Ibid.*

96. *Ibid.* However, recent studies have demonstrated that slitting the throat does cause pain, with a pain response in calves being detected for up to two minutes, and thus may be considered inhumane. Tim Edwards, 2009, "A study proving Jewish and Islamic methods of slaughtering animals are painful has led to renewed calls for a ban in Britain," *FirstPost*. Available from: http://www.thefirstpost.co.uk/54850,news-comment,news-politics,after-scientific-proof-of-pain-should-we-ban-islamic-and-jewish-religious-slaughter [accessed 9.16.2010].

97. 374 F. Supp. at 1290 n.8.

98. *Shackling and Hoisting,* YD 6.2000 (Sept. 20, 2000).

99. Congress, under the Establishment Clause, may accommodate religious practice, so long as the statute has a secular purpose, its primary effect neither advances nor inhibits religion, and it does not foster excessive government entanglement with religion. *Ibid.* at 1293 (citing *Lemon* v. *Kurtzman*, 403 U.S. 602, 612–613 (1971)).

100. Temple Grandin and Gary Smith, 2004, *Animal Welfare and Humane Slaughter.* Available from: http://www.grandin.com/references/humane.slaughter.html; see also Temple Grandin, *Recommended Ritual Slaughter Practices.* Available from: http://www.grandin.com/ritual/rec.ritual.slaughter.html.

101. The court also held that since the statute does not coerce conduct that infringes the plaintiffs' religious practice, there is no violation of the exercise clause. 374 F. Supp at 1294 ("By making it possible for those who wish to eat ritually acceptable meat to slaughter the animal in accordance with the tenets of their faith, Congress neither established the tenets of that faith nor interfered with the exercise of any other.").

102. 50 C.F.R. § 22.22(c).

103. *United States* v. *Vasquez-Ramos*, 522 F.3d 914, 918 (9th Cir. 2008) (quoting Public Laws, June 8, 1940, ch. 278, pmbl., 54 Stat. 250 (1940)).

104. *Ibid.* (quoting *United States* v. *Hardman*, 297 F.3d 1116, 1128 (10th Cir. 2002 (en banc))).

105. *New Jersey Society for the Prevention of Cruelty to Animals, et al.* v. *New Jersey Department of Agriculture, et al.*, 2007 WL 486764 (Feb. 16, 2007) (unpublished).

106. *Ibid.* at *3.

107. *Ibid.* at *5.

108. *Ibid.* at *1–2.
109. *New Jersey Soc. for Prevention of Cruelty to Animals v. New Jersey Dept. of Agriculture*, 196 N.J. 366, 955 A.2d 886 (NJ 2008) ("The facial challenge to the regulations in their entirety is rejected. The specific challenges to the reliance on 'routine husbandry practices' as defined in the regulations, and to the reliance on 'knowledgeable individual and in such a way as to minimize pain' are sustained. The specific challenges to the practices, with the exception of the practice of tail docking, are otherwise rejected.").
110. *Noah v. The Attorney General, et al.*, HCJ 9232/01, 215 (Israeli Supreme Court Aug. 11, 2003).
111. See generally M. Sullivan and D. Wolfson, 2007, "What's Good for the Goose ... The Israeli Supreme Court, Foie Gras, and the Future of Farmed Animals in the United States," 70-WTR *Law & Contemp. Probs.* 139; M. Sullivan and D. Wolfson, 2008, "The Regulation of Common Farming Practices," in *Animal Law and the Courts: A Reader*, 78–131, Thomson West, Eagan.
112. Justice Rivlin, siding with Justice Strasberg-Cohen, summarized his position as follows: "As for myself, I have no doubt that wild animals and house pets alike have feelings. They possess a soul that experiences the feelings of happiness and grief, joy and sorrow, affection and fear. Some develop feelings of affection toward their friend-enemy, man. Not all would agree with this view. All would agree, however, that these creatures feel the pain inflicted upon them by physical injury or by violent intrusion into their bodies. Indeed, one could justify the force-feeding of geese by pointing to the livelihood of those who raise geese and the gastronomical pleasure of others. Indeed, those wishing to justify the practice might paraphrase Job 5:7 [65]: It is right that man's welfare shall soar, even at the price of troubling birds of light. Except that it has a price – and the price is the degradation of man's own dignity." *Noah*, HCJ 9232/01 at 272.

Justice A. Grunis, the third Justice on the panel, finding that there was no alternative means to produce foie gras, concluded: "At the end of the day I have found that the force-feeding process does indeed cause suffering to the geese. And yet, in my opinion, it is unjustified to prevent the suffering of the geese by bringing suffering upon the farmers – which would be the result of their livelihood being wiped out in an instant." *Ibid.* at 249.
113. *Ibid.* at 250–251.
114. *NJSPCA*, 2007 WL 486764 at *3.
115. *Ibid.*
116. *Ibid.*
117. *Ibid.*
118. *Ibid.* at *4.
119. *Noah*, HCJ 9232/01 at 269–270.
120. *Ibid.* at 270.
121. *Ibid.* at 262.
122. *NJASPCA*, 2007 WL 486764 at *15.
123. *Ibid.* at *12.
124. See Pamela Frasch et al., 1999, "State Animal Anti-Cruelty Statutes: An Overview," 5 *Animal L.* 69, 77 & n.34 (noting that 30 states provide exemptions for accepted husbandry practices).
125. *NJASPCA*, 2007 WL 486764 at *3 (emphasis added).

126. *Ibid.*
127. *Ibid.* at *16-17.
128. *Ibid.* at *16.
129. *Ibid.*
130. *Ibid.* The NJ Supreme Court upheld this finding, stating in footnote 14: "Petitioners also included in their brief an assertion that force-feeding of geese for the purpose of creating foie gras is not humane. They suggest that because the regulations do not specifically prohibit this practice, the regulations can be interpreted to permit a practice that is inhumane and are, therefore, defective. We perceive of this as an application to this Court to impose a ban on this particular practice, a request that would be inconsistent with the organizational structure of the statute and the regulations and, in our view, a request that is more properly within the province of the legislative branch." *NJSPCA*, 955 A.2d at 908 n.14.
131. *NJASPCA*, 2007 WL 486764 at *7.
132. *Ibid.*
133. *Ibid.* at *7.
134. *Ibid.* at *8.
135. *Ibid.*
136. *Ibid.* at *9.
137. *Ibid.* at *9–10.
138. *Ibid.*
139. The NJ Supreme Court found this improper, stating: "Rather than creating a series of regulations that permit or disallow practices in accordance with whether they are humane, and rather than permitting practices only if performed in a specified manner, the agency instead authorized the practices in general and defined them as being humane by implicitly redefining humane itself. That is, the agency authorized the practices if performed by a knowledgeable person so as to minimize pain and equated that otherwise undefined person's choices with humane. This, however, has resulted in a regulation that is entirely circular in its logic, for it bases the definition of humane solely on the identity of the person performing the task, while creating the definition of that identity by using an undefined category of individuals of no discernable skill or experience." 955 A.2d at 913.
140. *Ibid.* at 910.
141. *Ibid.*
142. *Ibid.*
143. *Ibid.*
144. *NJASPC*, 2007 WL 486764 at *6.
145. *Noah*, HCJ 9232/01 at 256.
146. *Ibid.* at 262.
147. *Ibid.*
148. Interestingly, the only European report related to the study of animal protection cited by the New Jersey court was from 1976. *See NJASPC*, 2007 WL 486764 at *7 n.8.
149. *Noah*, HCJ 9232/01 at 263.
150. *Ibid.* at 265.
151. *Ibid.* at 259.
152. *Ibid.* at 267.

153. *Ibid.* at 268.
154. *Ibid.* at 271.
155. Sullivan and Wolfson, "What's Good for the Goose," at 152.
156. 16 U.S.C. § 3371.
157. 26 Conn. Gen. Stat. § 26-1.
158. Note Chad West, 2006, "Economics and Ethics in the Genetic Engineering of Animals," 19 *Harv. J. L. & Tech.* 413, 432.
159. S. White, 2009, "Animals in the Wild, Animal Welfare and the Law," in Sankoff and White eds., *Animal Law in Australasia*, 238 (quoting S.R. Harrop, "The Dynamics of Wild Animal Welfare Law," 9 *J. Envt'l L.* 287 (1997)).
160. Frasch et al., "State Animal Anti-Cruelty Statutes," at 77.
161. S. Waisman et al., eds., 2008, *Animal Law: Cases and Materials,* 3rd edn, at 474, Carolina Academic Press, Durham, NC.
162. *New Mexico* v. *Cleve*, 980 P.2d 23, 25 (New Mex. 1999).
163. *Ibid.* at 25–26.
164. *Ibid.* at 25.
165. *Ibid.* at 26 (quoting N.M. stat. § 30-18-1 (emphasis added)). Current N.M. stat. § 30-18-1B reads:

    B. Cruelty to animals consists of a person:
    (1) negligently mistreating, injuring, killing without lawful justification or tormenting an animal; or
    (2) abandoning or failing to provide necessary sustenance to an animal under that person's custody or control.

166. Section 18-2 prohibits the injury of "any animal or domesticated fowl which is the property of another," Section 18-3 prohibits the unlawful branding of "any animal which is the property of another;" Section 18-4 prohibits the "unlawful disposition of an animal," including "abandoning any livestock," taking livestock for use without the owner's consent, and "driving or leading any animal being the property of another from its usual range, without the consent of the owner." *Cleve*, 980 P.2d at 28.
167. *Ibid.* at 35.
168. *Ibid.* at 36 (quoting N.M.S.A. 1978 § 17-2-1 (1983)).
169. *Ibid.*
170. 1997 WL 187133 (Conn. Super 1997) (unpublished decision).
171. The regulations provide that "[n]uisance wildlife control operators may use cage traps, box traps, padded leg-hold traps in the burrow of a wild animal, other nonlethal methods, or shooting to alleviate nuisance situations caused by ... raccoons." Moreover, "nuisance wildlife control officer is encouraged to use nonlethal control practices, including relocation, for most nuisance species. However, the Commissioner has banned relocation of raccoons by control officers during 1996 due to the continued threat of rabies. Since defendant had trapped raccoons, he was required to destroy them within 24 hours as a condition of his license." *Ibid.* at *1.
172. The statute, Gen. Stats § 53-247(a), creates criminal liability for "(a) Any person who overdrives, drives when overloaded, overworks, tortures, deprives of necessary sustenance, mutilates or cruelly beats or kills or unjustifiably injures any animal ..." General Statutes § 108(a) defines "animal" to include "all brute creatures and birds." *Ibid.*

173. *Ibid.* at *2.
174. *Ibid.* at *3.
175. *Ibid.* at *4.
176. *Ibid.* at *4.
177. 648 N.E.2d. 1174 (Ct. App. Ind. 1995).
178. The seven guilty charges were: Criminal Recklessness, a Class B misdemeanor, two counts of Cruelty to an Animal, Class A misdemeanors, two counts of Illegal Taking of Migratory Waterfowl, Class C misdemeanors, and two counts of Illegal Possession of Migratory Waterfowl, Class C misdemeanors. The dissent, with the agreement of the majority stated: "This cause demonstrates the practice of 'overcharging.' The charging of eleven criminal offenses for the taking of two Canada geese is beyond the comprehension of any fair minded person. This practice should be condemned." *Ibid.* at 1182 n.5.
179. *Ibid.* at 1178.
180. See generally Aaron Lake, 1998, "1998 Legislative Review," 5 *Animal L.* 89, 95–99.
181. "Live traps and common rat and mouse traps, were excluded, along with the use of padded-jaw leg-hold traps in the 'extraordinary case' where it is 'the only method available to protect human safety.'" *Ibid.* at 97. The violation carries a misdemeanor penalty with fines of $300–$2,000 and up to one year in county jail.
182. The law also would have outlawed "possessing, purchasing, or selling the skin of a wolf known by the person to have been caught with the use of a snare." *Ibid.* at 95.
183. See for example 2009, *U.S. Crossbow Hunting Regulations* [online], Hunter's Friend. Available from: http://www.huntersfriend.com/crossbows/crossbow-state-regulations.htm [accessed: 5.25.2010].
184. *Report on Bowhunting* [online], Animals Rights Coalition. Available from: http://www.animalrightscoalition.com/doc/bowhunting_report.pdf [accessed: 5.25.2010].
185. *Ibid.*
186. Lake, "1998 Legislative Review," at 101 (quoting Minn. Const art. XIII, § 12).
187. *Ibid.* 100–101 (quoting Doug Smith, *Animal Rights Groups Target Constitutional Amendment*, Star-Tribune, Sept. 26, 1998, at 1B).
188. White, "Animals in the Wild, Animal Welfare and the Law," at 230. Thanks to Steven White for his summary from which this brief description is taken.
189. *Ibid.* at 238–239.
190. *Ibid.* at 239–240 (citing *Animal Liberation Ltd* v. *Dept Environment & Conservation* [2007] NSWSC 221 at [4]).
191. The Commonwealth has adopted Model Codes of Practice and the individual states have variously adopted their own codes of practice. In New Zealand, the National Animal Welfare Advisory Committee develops Codes of Welfare that are issued as regulations by the Ministry of Agriculture and Forestry. *Ibid.* at 241.
192. *Ibid.* at 242.
193. "Young-at-foot" is the term for young kangaroos who are able to live outside the mother's pouch.
194. "Joeys" are kangaroo babies living inside the mother's pouch.

195. White, "Animals in the Wild, Animal Welfare and the Law," at 242 (quoting [2008] AATA 717 at [50]).
196. *Ibid.* at 250 (quoting [2008] AATA 717 at [51]).
197. There are two: one for commercial hunting and one for non-commercial hunting. While there are some differences, especially in the firearms allowed, the general provisions are quite similar. See http://www.environment.gov.au/biodiversity/trade-use/wild-harvest/kangaroo/pubs/code-of-conduct-commercial.pdf (commercial); http://www.environment.gov.au/biodiversity/trade-use/wild-harvest/kangaroo/pubs/code-of-conduct-non-commercial.pdf (non-commercial).
198. Australia National Code of Practice for the Humane Shooting of Kangaroos and Wallabies for Commercial Purposes 2008, § 1.1, Available from: http://www.environment.gov.au/biodiversity/trade-use/wild-harvest/kangaroo/pubs/code-of-conduct-commercial.pdf [hereinafter Kangaroo Commercial Code].
199. Kangaroos and wallabies are both marsupials. The major differences between the two animals are their size (wallabies are smaller), their coloring (wallabies' coats tend to be brighter and with two to three different colors), and their teeth (wallabies live in bushy areas and eat mostly leaves, thus their teeth are flat to grind up the leaves; kangaroos mostly graze on grasses and thus have curved teeth with ridges for cutting and shearing the grass).
200. Kangaroo Commercial Code, § 2.4.
201. *Ibid.* § 1.3.
202. *Ibid.* § 2.4.
203. *Ibid.* § 5.1.
204. D. Thiriet, 2009, "Recreational Hunting – Regulation and Animal Welfare Concerns," in Sankoff and White, *Animal Law in Australasia*, 283.
205. *Ibid.* at 280.
206. See generally William Reppy, 2005, "Citizen Standing to Enforce Anti-Cruelty Laws by Obtaining Injunctions: The North Carolina Experience," 11 *Animal L.* 39.
207. N.C. Gen. Stat. 19A-2 (2003).

Chapter 3

1. Kevin Dolan, 2007, *Laboratory Animal Law*, 2nd edn, at 172, Blackwell, Oxford.
2. 7 U.S.C. § 2131.
3. 7 U.S.C. § 2132(f).
4. *Ibid.* § 2132(g).
5. H.R. Conf. Rep. No. 89-1848, at 6 (1966).
6. There are two groups of animals used in research: animals who are purpose-bred specifically for research and are provided to facilities by licensed Class A dealers and random-source animals obtained from individual owners, breeders, pounds and shelters many but not all of whom are obtained by facilities through Class B licensed dealers. See generally Committee on Scientific and Humane Issues in the Use of Random Source Dogs and Cats for Research, National Research Council, 2010, *Scientific and Humane Issues in the Use of Random-Source Dogs and Cats in Research*, National Academies

Press (examining the value of random-source animals in biomedical research and the role of Class B dealers who acquire and resell live dogs and cats to research and concluding that the government should obtain these animals from alternative sources because of the inadequate enforcement of animal welfare standards in regards to Class B dealers that in turn jeopardizes the welfare of these animals).

7. 7 U.S.C. § 2143.
8. The National Research Council maintains a "Guide For the Care and Use of Laboratory Animals" published by National Academies Press that has served as a basis for accreditation of institutions worldwide by the Association for Assessment and Accreditation of Laboratory Animal Care International. The eighth edition was released in the spring of 2010, integrating recently published data, scientific principles, and expert opinion to recommend practices for the humane care and use of animals in research, testing, and teaching. The new edition includes expanded coverage of the ethics of laboratory animal use; components of effective Animal Care and Use Programs; and new guidelines for the housing, environment, and enrichment of terrestrial and aquatic animals, and for veterinary and clinical care.
9. See for example 9 C.F.R. §§ 3.1 to 3.19 (Specifications for Dogs and Cats).
10. Henry Cohen, 2006, "The Animal Welfare Act," 2 *J Animal L.* 13; see also Mariann Sullivan, 2007, "The Animal Welfare Act – What's That?," 79-Aug *NY St. B.J.* 17.
11. 7 U.S.C. § 2132(g).
12. *ALDF v. Madigan*, 781 F. Supp. 797, 800 (D.D.C. 1992).
13. *Ibid*. at 800.
14. *Ibid*. at 802.
15. *Ibid*. at 802; see also Henry Cohen, 1987, "The Legality of the Agriculture Department's Exclusion of Rats and Mice from Coverage Under the Animal Welfare Act," 31 *St. Louis Univ. L.J.* 543, 545 n.19.
16. 781 F. Supp. at 803 n.4.
17. *Ibid*. at 805 n.5.
18. *ALDF v. Espy*, 23 F.3d 496 (DC Cir. 1994).
19. See *ALDF v. Glickman*, 154 F.3d 426 (DC Cir. 1998).
20. See Lauren Magnotti, 2006, "Giving a Voice to Those Who Can't Speak for Themselves: Toward Greater Regulation of Animal Experimentation," 13 *Buff. Envt'l L. J.* 179, 194–195.
21. 148 Cong. Rec. S612 (daily edn Feb. 12, 2002) (statement of Sen. Helms).
22. *Ibid*.
23. See 2002, *Animal Welfare Act May Not Protect All Critters* [online], USA Today. Available from: http://www.usatoday.com/news/nation/2002/05/07/animal-welfare.htm [accessed: 29.6.2010].
24. Magnotti, "Giving a Voice to Those Who Can't Speak for Themselves," at 195.
25. *ALDF v. Secretary of Agriculture*, 813 F. Supp 882, 886 (1993) (quoting 9 C.F.R. § 3.80(c)(1) (1992)).
26. *Ibid*. at 888 (quoting 9 C.F.R. § 3.81).
27. *Ibid*.
28. *Ibid*. at 887 n.8.
29. *Ibid*. at 887.

30. *Ibid.* at 889.
31. *Ibid.* at 887.
32. *ALDF* v. *Espy*, 29 F.3d 720 (D.C. Cir. 1994). Note that this case was brought before the D.C. case that revisited standing requirements discussed above. See note 19 above.
33. 813 F. Supp. at 890.
34. *Ibid.* at 888. Here are a few other notable quotes from Judge Richey's AWA cases. Note the refreshingly candid remarks and sincere concern for the welfare of animals: "In real world terms, if the veterinarian did what he or she was required by law, he or she might well be replaced or not paid." 813 F. Supp at 887.

    Quoting Judge Wright: "This appeal presents the recurring question which has plagued public regulation of industry: whether the regulatory agency is unduly oriented toward the interests of the industry it is designed to regulate, rather than the public interest it is designed to protect." *Ibid.* at 890 n.16.

    "This inertia on the part of the agency allows the mistreatment of birds, rats, and mice to continue unchecked by the agency charged with the protection of laboratory animals. The Court cannot believe this is what Congress had in mind." 781 F. Supp. at 805.

    "Because the case involves animals as opposed to human beings is not a legitimate excuse for such inordinate delay." 813 F. Supp at 890.
35. 9 C.F.R. §§ 2.30-2.37
36. See generally Darian Ibrahim, 2006, "Reduce, Refine, Replace: The Failure of the Three R's and the Future of Animal Experimentation," *U. Chi. Legal F.* 195. In 1959, scientists William Russell and Rex Burch proposed the 3Rs in W. Russell, R. Burch, 1958, *The Principles of Humane Experimental Technique* Methuen, London. Available from http://altweb.jhsph.edu/pubs/books/humane_exp/het-toc.
37. Ibrahim, "The Failure of the Three R's," at 205.
38. 7 U.S.C. § 2143(b).
39. Ibrahim, "The Failure of the Three R's," at 210.
40. 7 U.S.C. § 2143(6)(A)(ii).
41. 9 C.F.R. § 2.31.
42. Ibrahim, "The Failure of the Three R's," at 214.
43. *Ibid.* at 221–223.
44. 2005, *Audit Report: APHIS Animal Care Program, Inspection and Enforcement Activities* [online], USDA. Available from: http://www.usda.gov/oig/webdocs/33002-03-SF.pdf [accessed: 29.6.2010].
45. *Ibid.*
46. 2009, *USDA to Expand Access to Animal Research Data* [online], HSUS. Available from: http://www.hsus.org/animals_in_research/animals_in_research_news/usda_to_expand_access_to_animal_research_information_07102009.html [accessed: 29.6.2010].
47. 2007, *HSUS Files Suit Against USDA over FOIA Delays* [online], HSUS. Available from: http://www.hsus.org/animals_in_research/animals_in_research_news/hsus_files_suit_against_usda_over_foia_delays.html [accessed: 29.6.2010].
48. 2005, *USDA Agrees to Provide On-Line Animal Research Reports in Wake of HSUS Lawsuit* [online], HSUS. Available from: http://www.hsus.org/animals_in_

research/animals_in_research_news/usda_agrees_to_provide_online_reports. html [accessed: 29.6.2010].
49. 2008, *Lawsuit Reveals 81 Animal Research Facilities Not Meeting Animal Welfare Act Reporting Requirements* [online], HSUS. Available from: http://www.hsus. org/press_and_publications/press_releases/81_animal_research_facilities_ missing_animal_welfare_act_reports_032608.html [accessed: 29.6.2010].
50. 2009, *USDA to Expand Access to Animal Research Data* [online], HSUS. Available from: http://www.hsus.org/animals_in_research/animals_in_research_news/ usda_to_expand_access_to_animal_research_information_07102009.html [accessed: 29.6.2010]. See Annual Report of Research Facility (APHIS Form 7023), APHIS, Animal Welfare. Available from: http://www.aphis.usda.gov/ animal_welfare/efoia/7023.shtml [accessed: 3.7.2010].
51. Sullivan, "The Animal Welfare Act – What's That?," at 19.
52. *ALDF* v. *Espy*, 29 F.3d 720, 723 (1994).
53. 5 U.S.C. § 702.
54. 386 F.3d 1169 (9th Cir. 2004).
55. *Ibid.* at 1175–1176.
56. See generally Cass Sustein, 2000, "Standing for Animals (With Notes on Animal Rights)," 47 *UCLA L Rev.* 1333.
57. 5 U.S.C. §§ 551(2), 701(b)(2).
58. *ALDF* v. *Espy*, 23 F.3d 496 (D.C. Cir. 1994).
59. *Ibid.* at 500.
60. *Ibid.*
61. *Ibid.* at 504–505 (Williams, J. dissenting).
62. *Ibid.* at 501.
63. *Ibid.* at 501.
64. *Ibid.* (quoting *Lujan* v. *Defenders of Wildlife*, 112 S. Ct. 2130, 2145 (1992)).
65. *Ibid.* at 503.
66. *Ibid.*
67. *ALDF* v. *Espy*, 29 F.3d 720 (D.C. Cir. 1994).
68. *Ibid.* at 724.
69. *Ibid.* at 725.
70. *Ibid.* at 726 (Mikva, J. concurring).
71. *ALDF* v. *Glickman*, 943 F. Supp 44 (D. D.C. 1996).
72. *ALDF* v. *Glickman*, 154 F.3d 426 (D.C. Cir. 1998).
73. See *ibid.* at 429–431.
74. See *ibid.* at 431–438.
75. See *ibid.* at 438–444.
76. *Ibid.* at 447 (Sentelle, J. dissenting).
77. *Ibid.* at 449–450 (Sentelle, J. dissenting).
78. See *ibid.* at 451–453 (Sentelle, J. dissenting).
79. See *ibid.* at 454 (Sentelle, J. dissenting)
80. *Ibid.* at 444.
81. *Ibid.* at 444–445.
82. This brief summary is taken from Dolan, *Laboratory Animal Law*, an outstanding resource on this topic. For a summary of the laws in Australasia governing animals used in research see P. Gerber, 2009, "Scientific Experimentation on Animals: Are Australia and New Zealand Implementing the

3Rs?," in Peter Sankoff and Steven White, eds., at 212–229, *Animal Law in Australasia: A New Dialogue,* Federation Press, Sydney.
83. Dolan, *Laboratory Animal Law,* at 14; see generally *ibid.* ch. 3 ("The Protected Animal").
84. *Ibid.* at 34–35 (quoting Guidance App. F); see generally *ibid.* ch. 6 ("The Personal License").
85. *Ibid.* at 24 (quoting A(SP)A § 2(1)); see generally *ibid.* ch. 4 ("The Regulated Procedure").
86. *Ibid.* at 42; see generally *ibid.* ch. 7 ("The Project License").
87. *Ibid.* at 44 (quoting Directive 86/609/EEC, art. 23).
88. *Ibid.* at 52
89. *Ibid.*
90. *Ibid.* at 53 (quoting Guidance, app. I).
91. *Ibid.*
92. See *ibid.* at 55–58.
93. See generally *ibid.* ch. 9 ("Certificates of Designation").
94. Sullivan, "The Animal Welfare Act – What's That?," at 17.
95. 9 C.F.R. § 1.1.
96. 151 Cong. Rec. S6,031 (daily edn May 26, 2005) (statement of Sen. Santorum) ("Because the AWA only covers breeders and others who sell at wholesale, many puppy mill owners have successfully avoided AWA requirements by selling directly to the public.").
97. *What is a Puppy Mill?* [online], ASPCA. Available from: http://www.aspca.org/fight-animal-cruelty/puppy-mills/what-is-a-puppy-mill.html [accessed: 29.6.2010].
98. Va. Code Ann. § 3.2-6507.2

Commercial dog breeders shall:
1. Maintain no more than 50 dogs over the age of one year at any time for breeding purposes. However, a higher number of dogs may be allowed if approved by local ordinance after a public hearing. Any such ordinance may include additional requirements for commercial breeding operations;
2. Breed female dogs only: (i) after annual certification by a licensed veterinarian that the dog is in suitable health for breeding; (ii) after the dog has reached the age of 18 months; and (iii) if the dog has not yet reached the age of 8 years;
3. Dispose of dogs only by gift, sale, transfer, barter, or euthanasia by a licensed veterinarian;
4. Dispose of deceased dogs in accordance with § 3.2-6554;
5. Dispose of dog waste in accordance with state and federal laws and regulations; and
6. Maintain accurate records for at least five years including:
a. The date on which a dog enters the operation;
b. The person from whom the animal was purchased or obtained, including the address and phone number of such person;
c. A description of the animal, including the species, color, breed, sex, and approximate age and weight;
d. Any tattoo, microchip number, or other identification number carried by or appearing on the animal;

e. Each date that puppies were born to such animal and the number of puppies;

f. All medical care and vaccinations provided to the animal, including certifications required by a licensed veterinarian under this chapter; and

g. The disposition of each animal and the date.

Va. Code Ann. § 3.2-6507.5
Any commercial dog breeder violating any provision of this article is guilty of a Class 1 misdemeanor.

Va. Code Ann. § 3.2-6507.6
It shall be the duty of each attorney for the Commonwealth to enforce this article.

99. *Ibid.* § 3.2-6507.6.
100. 7 U.S.C. § 2132(h) (1970).
101. 543 F.2d 169 (D.C. Cir.1976).
102. *Ibid.* at 174.
103. *Ibid.* at 176 n.53.
104. *Ibid.*
105. *Ibid.* at 178.
106. *Ibid.*
107. 9 C.F.R. § 2.131.
108. See generally *Zoo Licensing Act 1981* [online], DEFRA, Available from: http://www.defra.gov.uk/wildlife-pets/zoos/licensing-act.htm [Accessed: 29.6.2010].
109. M. Radford, 2007, *Wild Animals in Travelling Circuses*, Circus Working Group; see also 2009, *Consultation on the use of wild animals in travelling circuses* [online], DEFRA. Available from: http://www.defra.gov.uk/corporate/consult/circus-wild-animals/consultation.pdf [accessed: 29.6.2010] (hereinafter DEFRA Consultation).
110. DEFRA, *The Keeping of Wild Animals*, Circular 1/2002 (emphasis added).
111. AWA 2009, § 9(1).
112. Radford, *Wild Animals in Travelling Circuses*, at 8.4.4.
113. DEFRA Consultation.
114. 2010, *Letter from Jim Fitzpatrick* [online], DEFRA. Available from: http://www.defra.gov.uk/corporate/consult/circus-wild-animals/jfitzpatrick-circus-letter.pdf [accessed: 29.6.2010].
115. See generally New Zealand Animal Welfare Act 1999, Wellington, MAF. Available from: http://www.legislation.govt.nz/act/public/1999/0142/latest/DLM49664.html?search=ts_act_Animal+Welfare+Act+1999_resel&sr=1 [accessed: 29.6.2010]; see also New Zealand Animal Welfare (Zoos) Code of Welfare 2004 [online]. Available from: http://www.wildlife.org.nz/zoocheck/zoo-code.pdf [accessed: 29.6.2010]; New Zealand Animal Welfare (Circuses) Code of Welfare 2005 [online]. Available from: http://www.biosecurity.govt.nz/files/regs/animal-welfare/req/codes/circus/circuses-code-of-welfare.pdf [accessed: 29.6.2010].
116. 278 F. Supp.2d 5 (D.D.C. 2003).
117. *Ibid.* at 24.
118. *Ibid.* at 25 n.4.
119. *Ibid.* at 27.

120. See generally *ASPCA* v. *Ringling Bros.*, Complaint for Declaratory and Injunctive relief, 2003 WL 24209908 (DDC).
121. 16 U.S.C. 1539(a)(1)(A).
122. "When used in reference to wildlife in captivity, 'enhance the propagation or survival ... includes but is not limited to the following activities when it can be shown that such activities would not be detrimental to the survival of wild or captive populations of the affected species:'
    (a) Provision of health care, management of populations by culling, contraception, euthanasia, grouping or handling of wildlife to control survivorship and reproduction, and similar normal practices of animal husbandry needed to maintain captive populations that are self-sustaining and that possess as much genetic vitality as possible;
    (b) Accumulation and holding of living wildlife that is not immediately needed or suitable for propagative or scientific purposes, and the transfer of such wildlife between persons in order to relieve crowding or other problems hindering the propagation or survival of the captive population at the location from which the wildlife would be removed; and
    (c) Exhibition of living wildlife in a manner designed to educate the public about the ecological role and conservation needs of the affected species."
    ASPCA v. *Ringling Bros.*, 502 F. Supp. 2d 103, 111 (D.D.C. 2007) (quoting 50 C.F.R. § 17.3).
123. 50 C.F.R. § 13.41.
124. *Ibid.* § 13.48.
125. 9 C.F.R. § 2.131(a),(b).
126. "Harass ... means an intentional or negligent act or omission which creates the likelihood of injury to wildlife by annoying it to such an extent as to significantly disrupt normal behavioral patterns which include, but are not limited to, breeding, feeding, or sheltering. This definition, when applied to captive wildlife, does not include generally accepted:
    (1) Animal husbandry practices that meet or exceed the minimum standards for facilities and care under the Animal Welfare Act,
    (2) Breeding procedures, or
    (3) Provisions of veterinary care for confining, tranquilizing, or anesthetizing, when such practices, procedures, or provisions are not likely to result in injury to the wildlife." 50 C.F.R. § 17.3.
127. "Harm ... means an act which actually kills or injures wildlife. Such act may include significant habitat modification or degradation where it actually kills or injures wildlife by significantly impairing essential behavioral patterns, including breeding, feeding or sheltering." *Ibid.*
128. See generally *ASPCA* v. *Ringling Bros.*, Answer, 2003 WL 24209909 (D.D.C.).
129. The United States has very liberal discovery rules to allow parties to obtain evidence in support of their respective cases. See generally Fed Rules of Civ Proc. 26-37. This often leads to serious arguments over relevance, privilege, and confidentiality of the materials requested. In this case the parties filed scores of discovery-related motions between 2003 and 2009. Unfortunately, parties often use discovery as a means to "wear down" the opposing party. This can be especially difficult for parties with relatively few resources (like the non-profit plaintiffs in this case) litigating against a party with substantial resources (like Feld Entertainment in this case).

130. *ASPCA* v. *Ringling Bros*, 502 F. Supp. 2d 103 (D.D.C. 2007).
131. *ASPCA* v. *FEI*, Defendant FEI's Response to Plaintiff's Post-Trial Brief, 2009 WL 1430653 at 3 (D.D.C.) (hereinafter FEI's Response).
132. *ASPCA* v. *FEI*, Plaintiff's Post-Trial Brief, 2009 WL 1266232 at 7 (D.D.C.) (hereinafter Plaintiff's Post-Trial Brief).
133. *Ibid.* at 8 (citing *In re* John F. Cuneo, AWA Docket No. 03-0023, Decision and Order as to James G. Zajicek (May 2, 2006), affirming Chief ALJ Decision as to James G. Zajicek (Aug. 17, 2005) (USDA concluding that striking an Asian elephant with a guide and creating a body wound is not a violation of the AWA)).
134. Plaintiff's Post-Trial Brief.
135. *Ibid.* Defendant countered that this amendment must have been an inadvertent mistake by Congress because no legislative history mentions removing "take" from the exempt pre-Act animals. FEI's Response, at 4.
136. Plaintiff's Post-Trial Brief, at 7.
137. *ASPCA* v. *FEI*, Plaintiff's Brief in Opposition to Post-Trial Brief of FEI, 2009 WL 1430652 at 4 (D.D.C.)
138. *Ibid.* at 8.
139. Plaintiff's Post-Trial Brief, at 8.
140. See generally *ASPCA* v. *FEI*, Plaintiff's Pre-Trial Brief, 2008 WL 5644793 (DDC). But see *ASPCA* v. *FEI*, Defendant's Pre-trial Brief, 2008 WL 5644794 (DDC).
141. D.Q. Wilbur, 2009, *Judge Rules in Favor of Circus in Lawsuit over Treatment of Asian Elephants* [online], Washington Post. Available from: http://www.washingtonpost.com/wp-dyn/content/article/2009/12/31/AR2009123101147.html?referrer=emailarticle [accessed: 30.6.2010].
142. *Ibid.*
143. *Ibid.*
144. Gaverick Matheny and Cheryl Leahy, 2007, "Farm-Animal Welfare, Legislation, and Trade," 70-WTR *Law & Contemp. Probs*, 325, 325.
145. D. Wolfson and M. Sullivan, 2004, "'Foxes in the Hen House': Animals, Agribusiness, and the Law: A Modern American Fable," in C. Sunstein and M. Nussbaum, eds., *Animal Rights: Current Debates and New Directions*, at 205–206, Oxford University Press, New York; see also David Wolfson, 1996, "Beyond the Law: Agribusiness and the Systemic Abuse of Animals raised for Food or Food Production," 2 *Animal L.* 123; David Wolfson, 1999, "McLibel," 5 *Animal L.* 21.
146. Further, recall from the cruelty chapter that many state anti-cruelty laws either exempt animals used for foods altogether or exempt "customary" farming practices which would clearly be violations of the cruelty statute if practiced on cats or dogs (or any other animal). Moreover, since they only prohibit "unnecessary" cruelty, harming animals in the production of food is likely considered "necessary" under most state laws. See Chapter 2; also Matheny and Leahy, "Farm-Animal Welfare, Legislation, and Trade," at 336.
147. 2010, *What is a CAFO?* [online], EPA. Available from: http://www.epa.gov/Region7/water/cafo/index.htm [accessed: 29.6.2010].
148. *Ibid.*

149. *Regulatory Definitions of Large CAFOs, Medium CAFO and Small CAFOs* [online], EPA. Available from: http://www.epa.gov/npdes/pubs/sector_table.pdf [accessed: 29.6.2010]. Hereinafter Factoryfarming.com.

150. See generally *Factoryfarming.com* [online], Farmsanctuary. Available from: http://farmsanctuary.org/issues/factoryfarming [accessed: 29.6.2010]; *Factory Farming Facts* [online], Idausa. Available from: http://www.idausa.org/facts/factoryfarmfacts.html [accessed: 29.6.2010].

151. A. Bruenjes, 2010, *Incentivizing Animal Cruelty: How Changing Federal Subsidies can Improve the Lives of Farmed Animals* (unpublished paper) (quoting W. Eubanks, III, 2009, "A Rotten System: Subsidizing Environmental Degredation & Poor Public Health with Our Nation's Tax Dollars," 28 *Stanford Envt'l L.J.* 213, 219, (citing D. Imhoff, 2007, *Food Fight: The Citizen's Guide to a Food and Farm Bill* at 22).

152. *Ibid.* (citing *Farm Subsidy Database* [online], Environmental Working Group. Available from http://farm.ewg.org/farm/region.php?fips=00000&progcode=total (corn subsidies totaled $56 billion compared to the next crop, wheat receiving $22 billion)) [accessed 29.6.2010].

153. *Ibid.* (quoting, 2004, "Challenging Concentration of Control in the Meat Industry," 117 *Harv. L. Rev.* 2643).

154. 7 U.S.C. § 1901.

155. H.R. Rep No 95-1336, at 3 (1978).

156. *Ibid.* at 5.

157. *Healthy Eating Pyramid* [online], Harvard School of Public Health. Available from: http://www.hsph.harvard.edu/nutritionsource/what-should-you-eat/pyramid/ [accessed: 29.6.2010].

158. 7 U.S.C. § 1902.

159. 49 U.S.C. § 80502 (a)(1).

160. *Ibid.* (b).

161. 49 U.S.C. § 80502 (a)(1).

162. *Ibid.* (a)(2)

163. *Ibid.*

164. 9 C.F.R. §§ 313.1 to 313.50

165. 9 C.F.R. § 89.1.

166. Matheny and Leahy, "Farm-Animal Welfare, Legislation, and Trade," at 335.

167. *Clay* v. *N.Y. Cent RR*, 231 N.Y.S. 424, 428 (N.Y. App. Div. 1928) (The Twenty-Eight Hour Law "does not apply to poultry; birds are not animals.").

168. Matheny and Leahy, "Farm-Animal Welfare, Legislation, and Trade," at 326; see also Jennifer Mariucci, 2008, "The Humane Methods of Slaughter Act: Deficiencies and Proposed Amendments," 4 *J. Animal L.* 149.

169. *Levine* v. *Connor*, 540 F. Supp. 2d 1113 (N.D. Cal. 2008).

170. 7 U.S.C. § 1902.

171. *Webster's International Dictionary of the English Language* 1446 (2d edn 1957).

172. *The American College Dictionary* 713 (1957).

173. "Dictionary definitions, however, change over time. *Compare* Black's Law Dictionary 1083 (4th edn 1951) (defining livestock as '[d]omestic animals used or raised on a farm') *with* Black's Law Dictionary 953 (8th edn 2004) (defining livestock as '[d]omestic animals and fowls that are kept for profit or pleasure')." 540 F. Supp. 2d at 1116 n.1.

174. See for example "Report on Changes in Farm Production and Efficiency," *USDA Statistical Bulletin* No. 233 (August 1958) (report included information about poultry when discussing livestock data); *Robinson* v. *Solano County*, 278 F.3d 1007, 1010 (9th Cir. 2002) (Including poultry), *State* v. *Nelson*, 499 N.W.2d 512, 514 (Minn. Ct. App. 1993) (excluding poultry); 7 U.S.C. § 182(4) (Chapter 9, entitled "Packers and Stockyards" and enacted in 1921, defines livestock as "cattle, sheep, swine, horses, mules, or goats."); 7 U.S.C. § 1523(b)(1) (the Federal Crop Insurance Act defines livestock to include "cattle, sheep, swine, goats, and poultry").
175. 540 F. Supp. 2d at 1117.
176. 21 U.S.C. §§ 601 *et seq.*
177. 21 U.S.C. § 451 ("It is essential in the public interest that the health and welfare of consumers be protected by assuring that poultry products distributed to them are wholesome, not adulterated, and properly marked, labeled, and packaged.").
178. 540 F. Supp. 2d at 1120–1121.
179. *Ibid.* at 1116 n.4.
180. See generally *FactoryFarming.com*.
181. See generally *Cruel Confinement* [online], HSUS. Available from: http://www.hsus.org/farm/camp/totc [accessed: 29.6.2010].
182. D. Winders, 2009, "Ohio's Issue 2 or, What is a Livestock Care Standards Board Anyway?," *ABA TIPS Animal Law Committee Newsletter*, Fall, 10.
183. C. Torres, 2009, *Ohio Voters Favor Livestock Standards Board* [online], Lancaster Farming. Available from: http://www.lancasterfarming.com/node/2370 [accessed: 29.6.2010].
184. 128th General Assembly, Ohio, Joint Resolution. Available at: http://www.legislature.state.oh.us/res.cfm?ID=128_SJR_6.
185. Winders, "Ohio's Issue 2" (quoting Joint Resolution Proposing to Enact Section 1 of Art. XIV of the Constitution of the State of Ohio, 128th Gen. Ass'y (Ohio 2009). Available from: http://www.sos.state.oh.us/sos/upload/ballotboard/2009/2-text.pdf [accessed: 29.6.2010]).
186. See A.T. Eliseuson and M.E.B. Norton, 2007, "The Chicago Foie Gras Ban: Innovative Milestone or Wild Goose Chase?," *ABA TIPS Animal Law Committee Newsletter*, Spring, 16.
187. See A.T. Eliseuson and M.E.B. Norton, 2007, "Constitutional Challenge Flies South: The Court Upholds Windy City Foie Gras Ban," *ABA TIPS Animal Law Committee Newsletter*, Fall, 13.
188. M. Davey, 2008, *Ban Lifted, Foie Gras Is Back on the Menu in Chicago* [online], New York Times. Available from: http://www.nytimes.com/2008/05/15/us/15liver.html [accessed: 29.6.2010].
189. See generally Compassion Over Killing [online], *Campaigns*. Available from: http://www.cok.net/camp/ [accessed: 3.7.2010].
190. See for example 21 U.S.C. § 601(n)(1) (Federal Meat Inspection Act prohibiting misleading labeling of meat products); 21 U.S.C. § 453 (h)(1) (Poultry Products Inspection Act prohibiting misleading labeling of poultry products).
191. See generally *COK Petitions FDA to Mandate Full Disclosure on Egg Cartons* [online], Compassion Over Killing. Available from: http://www.cok.net/camp/egg_labeling/ [accessed: 29.6.2010].

192. *Citizens's Petition to Change the Labeling Requirements for Eggs Sold in the US* [online], Compassion Over Killing, at 2–3. Available from: http://www.cok.net/files/FDA_egg_labeling_petition.pdf [accessed: 30.6.2010] (citing 2000, Zogby International Poll, and 2004, Golin/Harris International study) [hereinafter COK Petition].

193. *Meat and Poultry Labeling Terms* [online], USDA, Available from: http://www.fsis.usda.gov/FactSheets/Meat_&_Poultry_Labeling_Terms/index.asp [accessed: 29.6.2010].

194. See *Alliance for Bio-Integrity* v. *Shalala*, 116 F. Supp.2d 166, 178 (D.D.C. 2000).

195. See Council Directive 1999/74/EC of 19 July 1999, Article 5(2); Council Regulation (EC) No. 1028/2006 on marking standards for eggs, art. 4 (June 19, 2006). Available from: http://eur-lex.europa.eu/LexUriServ/LexUriServ.do?uri=OJ:L:2006:186:0001:0005:EN:PDF [accessed: 29.6.2010].

196. This section is drawn largely from the 2009 report, *Farm Animal Welfare in Great Britain: Past, Present and Future*, Farm Animal Welfare Council (hereinafter FAWC report). The Farm Animal Welfare Council was established in 1979 to review and study the welfare of animals used for food in the United Kingdom, advise the government of any needed legislation, communicate with outside bodies, the European Commission, and the public, and publish its advice independently. *Ibid.* at 1.

197. Ruth Harrison was one of the "movers and shakers" of the modern era animal movement who led a long crusade against factory farming. She died in June 2009. The Animal Welfare Institute published a tribute to her at http://www.awionline.org/ht/d/ContentDetails/id/2128/pid/2535.

198. FAWC report, at 6.

199. *Ibid.* at 12 n.25 (emphasis added).

200. *Ibid.* at 9.

201. Council Regulation (EC) No1/2005 art. 3 [online]. Available from: http://eur-lex.europa.eu/LexUriServ/site/en/oj/2005/l_003/l_00320050105en00010044.pdf [accessed: 29.6.2010] (emphasis added).

202. *Ibid.*

203. See generally *Animal Health and Welfare, Animal Welfare on the Farm* [online], EUROPA. Available from: http://ec.europa.eu/food/animal/welfare/farm/index_en.htm [accessed: 29.6.2010].

204. FAWC report, at 1.

205. The Five Freedoms are:
Freedom from hunger and thirst, by ready access to water and a diet to maintain health and vigor,
Freedom from discomfort, by providing an appropriate environment.
Freedom from pain, injury and disease, by prevention or rapid diagnosis and treatment.
Freedom to express normal behavior, by providing sufficient space, proper facilities and appropriate company of the animal's own kind.
Freedom from fear and distress, by ensuring conditions and treatment, which avoid mental suffering. *Ibid.* at 2.

206. See *Freedom Food* [online], RSPCA. Available from: http://www.rspca.org.uk/freedomfood [accessed: 29.6.2010].

207. FAWC report, at 45–46.

208. *Ibid.* at 45.

209. EU Welfare Quality Project website, http://www.welfarequality.net/everyone.
210. FAWC report, at 39.

Chapter 4

1. See for example D.C. Stat. § 8–1801 et seq. (ch. 18, "Animal Control").
2. City of Akron, ex rel. *Christman-Resch* v. *City of Akron*, 825 N.E.2d 189, 194 (Oh Ct App. 2005).
3. *Ibid.* at 196–197.
4. *Ibid.* at 198.
5. *Ibid.* at 198–199.
6. *Commonwealth of Penn.* v. *Creighton*, 639 A.2d 1296, 1297 n.1 (Penn. 1994) (quoting Borough of Carnegie Ordinance No. 1089).
7. *Ibid.* at 1301.
8. *Ibid.*
9. *Ibid.* at 1298.
10. *Holt* v. *Sauk Rapids*, 559 N.W.2d 444, 445–446 (Minn. Ct. App.1997).
11. *See Animal Shelter Euthanasia* [online], American Humane Association. Available from: http://www.americanhumane.org/about-us/newsroom/fact-sheets/animal-shelter-euthanasia.html [accessed: 16.6.2010].
12. *Ibid.*
13. *The Companion Animal Protection Act* [online], No Kill Advocacy Center. Available from: http://www.nokilladvocacycenter.org/capa.html [accessed: 16.6.2010] (hereinafter CAPA).
14. See for example Oreo's Law, N.Y. Bill A.9449D/S.6412D (Mar. 15, 2010) (The bill was voted down in June 2010. In essence, it would have required the release of a shelter animal to a rescue group upon request of the rescue group prior to euthanasia of the animal.) Oreo's Law was "named in memory of the pit bull mix who became well-known after she survived abuse at the hands of her former owner, including a fall from a six-story building." Although a no-kill animal sanctuary specializing in the care and rehabilitation of abused animals offered to take Oreo, the New York ASPCA refused their request and euthanized her after they determined she was untreatably aggressive. Press Release, The Assembly, State of New York, Albany (Nov. 18, 2009). Oreo's Law was patterned after a section of Hayden's Law, comprehensive legislation adopted by California in 1998 to promote a no-kill philosophy within their shelters. See generally T. Bryant, *Hayden Law* [online] Maddie's Fund. Available from: http://www.maddiesfund.org/Resource_Library/Hayden_Law.html [accessed: 4.7.2010].
15. ACR 74, *Animal Shelters: No Kill Movement Policies* (May 18, 2009) [online]. Available from: http://leginfo.ca.gov/pub/09-10/bill/asm/ab_0051-0100/acr_74_bill_20090518_introduced.pdf [accessed: 16.6.2010].
    In July 2010, Delaware amended its animal sheltering law and adopted several of the CAPA requirements designed to save homeless pets from unnecessary killing. "An Act to Amend Title 3 of the Delaware Code Relating to Animals Held in Shelter," Delaware Senate Bill No. 280, 145th General Assembly (2010).
16. See generally J. Schaffner, ed., 2009, *A Lawyer's Guide to Dangerous Dog Issues*, ch. 2, American Bar Association, Chicago.

17. The U.K. Dangerous Dog Act bans "any dog of the type known as the pit bull terrier, ... Japanese tosa, ... [or] being a type appearing ... to be bred for fighting or to have the characteristic of a type bred for that purpose." United Kingdom Dangerous Dogs Act 1991, ch. 65, § 1.

18. B. Klaassen, J.R. Buckley and A. Esmail, 1996, "Does the Dangerous Dogs Act Protect Against Animal Attacks: A Prospective Study of Mammalian Bites in the Accident and Emergency Department," 27 *Injury* 2, 89–91.

19. B. Rosado et al., 2007, "Spanish Dangerous Animals Act: Effect of the Epidemiology of Dog Bites," 2 *Journal of Veterinary Behavior* 5, 166–174.

20. See P. Marcus, 2009, *Do Dog Breed Bans Work?* [online], Denver Daily News. Available from: http://www.thedenverdailynews.com/article.php?aID=3473 [accessed: 16.6.2010].

21. 2002, *Report of the Vicious Animal Legislation Task Force* [online], Understand-a-bull. Available from: http://www.understand-a-bull.com/BSL/Research/PGCMD/PGCMTOC1.htm [accessed: 16.6.2010].

22. *Ibid.*

23. L. VanKavage and J. Dunham, 2010, "Fiscal Bite & Breed Discrimination: Utilizing Scientific Advances & Economic Tools in Lobbying," paper presented at Mid-Atlantic Animal Law Symposium [online] (9 April 2010). Available from: http://www.msba.org/sec_comm/sections/animallaw/docs/MidAtlanticAnimalLawfinallvjd.pdf.

24. J. Durham and Associates for the Best Friends Animal Society, *Best Friends Breed Discriminatory Legislation Fiscal Impact* [online], Best Friends Network. Available from: http://www.guerrillaeconomics.biz/bestfriends/ [accessed: 16.6.2010].

25. *Ibid.*

26. VanKavage and Dunham, "Fiscal Bite & Breed Discrimination," at 8.

27. See generally Schaffner, *Guide to Dangerous Dog Issues*, ch. 3.

28. See for example Ohio Revised Code § 955.22D (referring to "breed of dog commonly known as a Pit Bull").

29. 2006 WL 513946 at *3.

30. K. Delise, *Types of Dog Bites* [online], National Canine Research Council. Available from: http://nationalcanineresearchcouncil.com/dog-bites/types-of-dog-bites/ [Accessed: 16.6.2010]. Karen Delise, the researcher for this study, is a veterinary technician, author of *The Pitbull Placebo* and *Fatal Dog Attacks*, and founder and director of research for the National Canine Research Council (NCRC).

31. *Toledo* v. *Tellings*, 871 N.E.2d 1152 (Oh 2007).

32. *Dias* v. *Denver*, 567 F.3d 1169, 1184 (10th Cir. 2009).

33. *Ibid.* at 1184.

34. Dangerous Dogs Act 1991, ch. 65, § 3.

35. *Control of Dogs, the Law and You* [online], DEFRA. Available from: http://www.defra.gov.uk/wildlife-pets/pets/cruelty/documents/ddogslawyouleaflet.pdf [accessed: 16.6.2010] .

36. "Serious injury" means any physical injury that results in broken bones or lacerations requiring multiple sutures or cosmetic surgery. D.C. Stat. § 8-1901(6).

37. *Ibid.* § 8-1901(4)(A).

38. *Ibid.* § 8-1901(1)(A).

39. *Kirkham* v. *Will*, 311 Ill. App.3d 787 (App. Ill. 5th Dist. 2000).
40. *Ibid.* at 791 (citing *Nelson* v. *Lewis*, 36 Ill. App. 3d 130, 131 (1976)).
41. D.C. Stat. § 8-1904. "Proper enclosure" means secure confinement indoors or secure confinement outdoors in a locked structure designed and constructed to: (A) Deter escape of the dog; (B) Protect the dog from the elements; and (C) Prevent contact with the dog from humans and other domestic animals." *Ibid.* § 8-1901(5).
42. *Ibid.* § 8-1905.
43. *Ibid.* § 8-1906.
44. See generally Schaffner, *Guide to Dangerous Dog Issues*, at 18–22.
45. Nathan Winograd, 2005, *Do Feral Cats Have a Right to Live: A National No Kill Standard for Feral Cats*, 8 [online]. Available from: http://www.nokilladvocacycenter.org/pdf/Feral%20Cats.pdf [accessed: 8.21.2010].
46. Mitsuhiko Takahashi, 2004, "Cats v. Birds in Japan: How to Reconcile Wildlife Conservation and Animal Protection," 17 *Geo. Int'l Envt'l L. Rev.* 135, 149 (Fall) (quoting Wyo. Stat Ann § 11-31-301(a)).
47. See generally Verne R. Smith, 2009, "The Law and Feral Cats," 3 *J. Animal L. & Ethics* 7.
48. Prince George's County Code § 3-101(11).
49. *Ibid.* § 3-101 (Definitions).
50. *Ibid.* § 3-101(57).
51. *Ibid.* § 3-101(50).
52. *Ibid.* § 3-101(72).
53. *Ibid.* § 3-101(12).
54. *Ibid.* § 3-135(b).
55. *Ibid.* §§ 3-122, 3-140.
56. See generally CAPA; see also Illinois TNR law, 510 ILCS 5, et seq.
57. *TNR Cost Savings Calculator* [online], Best Friends Animal Society. Available from: http://www.guerrillaeconomics.biz/communitycats/ [accessed: 16.6.2010].
58. T. Longcore, C. Rich and L. Sullivan, 2009, "Critical Assessment of Claims Regarding Management of Feral Cats by Trap-Neuter-Return," 23 *Conservation Biology* 4, 887–894 [online]. Available from: http://ca.audubon.org/chapter_assets/Longcoreetal2009ConBio.pdf [accessed: 16.6.2010]; see also Shawn Gorman and Julie Levy, 2004, "Public Policy Toward the Management of Feral Cats," 2 *Pierce L. Rev.* 157.
59. In June 2008, a number of organizations, including several Audubon Societies in the Los Angeles, California area, filed a complaint against the City of Los Angeles requesting that the TNR efforts of the city be enjoined until an environmental review pursuant to the California Environmental Quality Act is completed. In November 2009, at a hearing before Judge McKnew, the City argued, inter alia, that neutering and returning homeless cats already living in the existing environment has no adverse effect on the environment. The judge granted a broad injunction ending all TNR activities and indicating that returning the cats to their colony would be bad for wildlife. *In re* the Matter of the *Urban Wildlands Group et al.* v. *City of Los Angeles*, Ruling on Petition for Writ of Mandamus, BS115483 (McKnew, J., Dec. 4, 2009).
60. W. Anderson and A. Vantiotis, 2008, "Animal v. Animals: A False Choice," *ABA TIPS Animal Law Committee Newsletter*, Spring, at 14 [online]. Available

from: http://www.alleycat.org/NetCommunity/Document.Doc?id=30 [accessed 16.6.2010].

61. *Ibid.* at 29.

62. *Ibid.* at 14.

63. See Florida Fish and Wildlife Conservation Commission. Available from: http://myfwc.com/ABOUT/index.htm [accessed: 16.6.2010].

64. Some animals do not require permits for ownership: for example, hedgehogs, prairie dogs, chinchillas, honey possums, squirrels, chipmunks, parrots, and certain reptiles. Also, "camels, llamas, wild horses, jungle fowl, common guinea fowl and peafowl are considered domestic/domesticated species and do not require a permit. Ratites and bison possessed for farming purposes do not require a permit." See *Captive Wildlife Categories* [online], Florida Fish and Wildlife Conservation Commission. Available from: http://myfwc.com/rulesandregs/Rules_CaptiveCategories.htm#concern [accessed: 16.6.2010].

65. Such animals include chimpanzees, gorillas, leopards, jaguars, bears, elephants, crocodiles, and komodo dragons. *Ibid.*

66. Howler monkeys, cheetahs, wolverines, ocelots, cougars, panthers, and ostrich. *Ibid.*

67. See also *Incidents Involving Captive Held Exotic Animals* [online], Born Free USA. Available from: http://www.bornfreeusa.org/popups/a3b_captive_animal_incidents.php [accessed: 16.6.2010].

68. See generally *Summary of State Laws Relating to Private Possession of Exotic Animals* [online], Born Free USA. Available from: http://www.bornfreeusa.org/b4a2_exotic_animals_summary.php [accessed: 16.6.2010]; http://www.bornfreeusa.org/b4a2_exotic_animals_map.php (20 states ban private ownership of all exotic animals, nine states ban private ownership of some exotic animals, twelve states require the owner to obtain a license or permit to own exotic animals, and nine states have no license requirements but may regulate some aspect of ownership).

69. See for example Fl. Stat § 379.101(27) (defining open season as "that portion of the year wherein the laws of Florida for the preservation of fish and game permit the taking of particular species of game or varieties of fish").

70. *Hunting* [online], National Shooting Sports Foundation. Available from: http://www.nssf.org/hunting/index.cfm?AoI=hunting [accessed: 16.6.2010].

71. 2009, *FWC Seeks Hunters to Complete Deer Rut Survey* [online], Florida Fish and Wildlife Conservation Commission. Available from: http://www.nssf.org/hunting/index.cfm?AoI=hunting [accessed: 16.6.2010].

72. See generally John Cooper, 2009, "Hunting as an Abusive Sub-culture," 302–316, and P. Cohn and A. Linzey, 2009, "Hunting as a Morally Suspect Activity," 317–328, both in A. Linzey, ed., *The Link between Animal Abuse and Human Violence*, Sussex Academic Press, Brighton/Portland.

73. D. Thiriet, 2009, "Recreational Hunting – Regulation and Animal Welfare Concerns," in Peter Sankoff and Steven White, eds., *Animal Law in Australasia: A New Dialogue*, at 259–260, Federation Press, Sydney.

74. See generally Katherine Hessler, "Where Do We Draw the Line Between Harassment and Free Speech? An Analysis of Hunter Harassment Law," 3 *Animal L.* 129 (1997).

75. 556 N.W.2d 578 (Minn. App. 1996).

76. *Ibid.* at 581 (quoting Minn. Stat.§ 97A.037 (1994)) (emphasis added).

77. *Ibid.* at 582.
78. *People* v. *Sanders*, 182 Ill.2d 524 (Ill. 1998).
79. 862 F.2d 432 (2d Cir. 1988).
80. *Ibid.* at 437.
81. Conn. Gen. Stat. §§ 53a–183a.
82. 796 A.2d 542 (Conn. 2002).
83. *Sanders*, 696 N.E.2d at 1149–1150 (Harrison, J. concurring in part and dissenting in part).
84. *Ibid.* at 1151–1152.
85. *Ball*, 796 A.2d at 553.
86. *Ibid.*
87. *Ibid.* at 554.
88. *Ibid.* at 554-55.
89. *Sanders*, 696 N.E.2d at 1152.
90. Thiriet, "Recreational Hunting," at 280.
91. *Ibid.*
92. See for example New York McKinney's ECL § 11-0524; N.H. Rev. Stat. § 210:24-b; Conn. Gen Stat. § 26-47.
93. 8 Natural Resources ch. 16, § 10-908 (MD).
94. Landowners wishing to control wildlife on their property need not have a permit if they are controlling unprotected species, including woodchucks, feral pigeons, house sparrows, and European starlings; or mice, moles, or rats when the animals are causing damage to property.
95. DC Wildlife Protection Act, B18-498 (2009). Available from: http://www.dccouncil.washington.dc.us/images/00001/20091021170826.pdf.
96. This discussion is derived from the work of Dr. Jay Kirkpatrick, Ph.D., Director of The Science and Conservation Center in Billings Montana. See for example J. Kirkpatrick and J. Turner Jr., 1997, "Urban Deer Contraception: The Seven Stages of Grief," 25 *Wildlife Society Bull.* 2, 515–519.
97. Response to PA Game Commission by Jay Kirkpatrick, Ph.D. (Jan. 2007) (citing Rutberg et al., 2004, 116 *BiologicalConservation* 243–250).
98. M. Newman, 2009, *Deer Immunocontraception* [online], NIST. Available from: http://www.nist.gov/public_affairs/factsheet/deer.cfm [accessed: 16.6.2010].
99. A. Rutberg and R. Naugle, 2008, "Deer-Vehicle Collision Trends at a Suburban Immunocontraception Site," 2 *Human-Wildlife Conflicts* 1, 60–67.
100. R. Naugle et al., 2002, "Field Testing Immunocontraception on White-Tailed Deer (Octocoileus virginiamus) on Fire Island National Seashore, New York, USA," *Reproduction Supplement* 60, 143–153.
101. John W. Turner, 2008, "Controlled-Release Components of PZP Contraceptive Vaccine Extend Duration of Infertility," 35 *Wildlife Research* 555–562, at 561.
102. Endangered Species Act, 16 U.S.C. §§ 1531–1543.
103. Migratory Bird Treaty Act, 16 U.S.C. §§ 703–712. (The Act was enacted to implement the International Convention for the Protection of Migratory Birds between the United States and Great Britain (acting for Canada). The Act prohibits the "killing, capturing or selling any of the migratory birds included in the terms of the treaty except as permitted by regulations" that are administered by the Department of the Interior.)
104. Wild Free-Roaming Horses and Burros Act, 16 U.S.C. § § 1331-1340 (The Act "was enacted to preserve and protect wild free-roaming horses and burros on

the public lands of the United States. The Act provides that the term 'wild free-roaming horses and burros' means all unbranded and unclaimed horses and burros on public lands of the United States." 38 C.J.S. Game § 35.)

105. Marine Mammal Protection Act, 16 U.S.C. § 1361 et seq. The Act imposes a moratorium upon the taking and import of marine mammals and products derived from them subject to exceptions under permits and regulations issued by the Secretary of the department in which the National Oceanic and Atmospheric Administration is operating (as to matters pertaining to animals of the order Cetacea and animals, other than walruses, of the order Pinnipedia), or by the Secretary of the Interior (as to matters pertaining to other marine mammals). The Act also contains an exception pertaining to the "taking of marine mammals by Indians, Aleuts, and Eskimos residing in Alaska for subsistence purposes or for the purpose of creating and selling authentic native articles of handicrafts and clothing." 35A Am. Jur. 2d Fish, Game, and Wildlife Conservation § 68. See generally J. Curnutt, 2001, *Animals and the Law: A Sourcebook*, ABL-C10 e-book, ch. 6.

106. Note Joe Mann, 1999, "Making Sense of the Endangered Species Act: A Human Centered Justification," 7 *NYU Envtl. L.J.* 246, 248.

107. The United States tied with the United Kingdom for 15th in the International Property Rights Alliance ranking of countries' level of property rights protection. New Zealand ranked 4th and Australia ranked 8th. The 2009 IPRI comprises a total of ten variables, which are divided into the three main components: Legal and Political Environment (LP), Physical Property Rights (PPR), and Intellectual Property Rights (IPR). The countries are ranked from the highest to the lowest score, representing how well a country protects property rights. Thus the country ranked first scored the highest. 2009, *International Property Rights Index* [online], Property Rights Alliance. Available from: http://www.internationalpropertyrightsindex.org/atr_Final1.pdf [accessed: 16.6.2010].

108. See generally P. Baldwin, E.H. Buck and M.L. Corn, 2005, CRS Report for Congress 2005, *The Endangered Species Act: A Primer*, Library of Congress, Washington, DC (hereinafter *ESA Primer*).

109. 16 U.S.C. § 1532(6).

110. 16 U.S.C. § 1532(20).

111. 16 U.S.C. § 1532(19).

112. 16 U.S.C. § 1538.

113. *ESA Primer*, at CRS-5 to 6.

114. Mann, "Making Sense of the Endangered Species Act," at 268.

115. *ESA Primer*, at CRS-7 (citing 16 U.S.C. § 1536(a)(2)).

116. *Ibid.* (As of March 30, 2005, recovery plans had been completed for 1,031 United States species. http://ecos.fws.gov/tess/html/boxscore.html).

117. *Ibid.* at CRS-9. The six members are: the Secretary of Agriculture, the Secretary of the Army, the Chairman of the Council of Economic Advisors, the Administrator of the Environmental Protection Agency, the Secretary of the Interior, and the Administrator of the National Oceanic and Atmospheric Administration. 16 U.S.C. § 1563(e)(3).

118. *ESA Primer*, at CRS-9.

119. *Ibid.* at CRS-10.

120. Mann, "Making Sense of the Endangered Species Act," at 247 (quoting Joseph Petulla, *American Environmentalism* 51 (1980)).
121. *Ibid.* (quoting 143 Cong Rec. E116 (daily edn June 4, 1997) (Congresswoman Helen Chenoweth, Chair of the House Subcommittee on Forests and Forest Health, in turn quoting former Sen. Malcolm Wallop). Mann quotes Rush Limbaugh, a conservative talk show host in the United States, as stating in response to issues surrounding the preservation of the northern spotted owl: "If the owl can't adapt to the superiority of humans, screw it." *Ibid.* (quoting Rush Limbaugh, *The Way Things Ought to Be* 160 (1992)).
122. See generally Mann, "Making Sense of the Endangered Species Act."
123. *Ibid.* at 252.
124. *Ibid.* at 253–254, 257, 259.
125. *Ibid.* at 255 (quoting Rep. Henry Helstoski).
126. *Ibid.* (quoting Rep. Leonor Sullivan).
127. *Ibid.* at 259 (quoting Sen. Tenney).
128. *Ibid.* at 263.
129. *Ibid.* at 266 (quoting Sen. Cranston).
130. 437 U.S. 153 (1978).
131. Once the exemption amendment passed, the Tellico Dam was exempted from the ESA, completed, and operating by 1980.
132. Mann, "Making Sense of the Endangered Species Act," at 260.
133. 16 U.S.C. § 1151–1159.
134. 16 U.S.C. § 668.
135. Mann, "Making Sense of the Endangered Species Act," at 273 (quoting H.R. Rep No. 89-2154 at 4 (1966)).
136. *Ibid.* (quoting BGEPA, ch. 278, 54 Stat. 250 (1940) (enacting clause)).
137. *HSUS* v. *Kemthorne*, 481 F. Supp. 2d 53 (2006).
138. 16 U.S.C. § 1539(a)(1)(A).
139. 50 C.F.R. § 17.22(a)(2).
140. 481 F. Supp. 2d at 58.
141. *Ibid.* at 62.
142. *Ibid.* at 63 n.4.
143. *Ibid.* at 63.
144. *Ibid.* at 63 (quoting Pls. Mem. Prelim. Inj., Ex. F (Tr. of Prelim. Inj. Hearing at 11, *Defenders of Wildlife* v. *Norton*, Civil Action No. 05-1573 (D.D.C.))).
145. *Ibid.* at 69.
146. *Ibid.*
147. *Ibid.* at 70.
148. *Ibid.* at 71.
149. *Ibid.*
150. *Ibid.* at 71–72.
151. *Defenders of Wildlife, et al.* v. *Hall*, 565 F.Supp.2d 1160 (D. Mont. 2008).
152. *Ibid.* at 1162.
153. *Ibid.* at 1163.
154. *Ibid.* at 1164.

Chapter 5

1. See J. Madison, 2004, "Animals are Not Constitutional Persons," in Andrew Linzey and Paul B. Clarke, eds., *Animal Rights: An Historical Anthology*, Columbia University Press, New York, 127–129 (discussing the status of slaves under the constitution and noting that "the slave may appear to be degraded from the human rank, and classed with those irrational animals which fall under the legal denomination of property").
2. *Viilo* v. *Eyre*, 547 F.3d 707 (7th Cir. 2008).
3. *Ibid.* at 709 (quoting *Harlow* v. *Fitzgerald*, 457 U.S. 800, 818 (1982)).
4. J. Diedrich, 2008, *Dog Owner Loses Civil Rights Case in Police Shooting of Pooch* [online], JSOnline. Available from: http://www.jsonline.com/news/milwaukee/35641019.html [accessed: 17.6.2010].
5. G. Bolliger, 2007, "Summary: Animal Welfare in Constitutions" (abstract from *Constitutional and Legislative Aspects of Animal Welfare in Europe*) [online]. Available from: http://www.tierimrecht.org/de/PDF_Files_gesammelt/Abstract_Bruessel_TIR_Papier.pdf [accessed 17.6.2010].
6. *Ibid.*
7. Fl. Const. art. IV, § 9.
8. See E. O'Carroll, 2008, *Ecuador Constitution would grant inalienable rights to nature*, Christian Science Monitor [online] Available from: http://www.csmonitor.com/Environment/Bright-Green/2008/0903/ecuador-constitution-would-grant-inalienable-rights-to-nature [accessed: 16.9.2010].
9. Bolliger, "Summary: Animal Welfare in Constitutions."
10. European Union, 2008, *Consolidated Version of the Treaty on the Functioning of the European Union* [online], art. 13. Available from: http://eur-lex.europa.eu/LexUriServ/LexUriServ.do?uri=OJ:C:2008:115:0047:0199:EN:PDF [17.6.2010] (emphasis added). Note that animals are also present in Article 36 as well, that states: "The provisions of Articles 34 and 35 shall not preclude prohibitions or restrictions on imports, exports or goods in transit justified on grounds of public morality, public policy or public security; the protection of health and life of humans, animals or plants; the protection of national treasures possessing artistic, historic or archaeological value; or the protection of industrial and commercial property. Such prohibitions or restrictions shall not, however, constitute a means of arbitrary discrimination or a disguised restriction on trade between Member States." *Ibid.* art. 36; see Case C-5/94, *The Queen* v. *Ministry of Agriculture, Fisheries, and Food, ex parte* Hedley Lomas (Ireland) Ltd., May 23, 1996, E.C.R. I-255, 3 *Colum. J. Eur. L.* 132, 140 ("Article 36 allows the maintenance of restrictions on the free movement of goods, if these are justified on grounds of the protection of the health and life of animals, which constitutes a fundamental requirement recognized by Community law. However, it is settled Community law that recourse to Article 36 is no longer possible where Community Directives provide for harmonization of the measures necessary to achieve the specific objective to be furthered by reliance upon this provision").
11. Recall from Chapter 1 that the EU does not have a constitution, but the Consolidated Treaties on the Functioning of the EU, last modified by the Lisbon Treaty, is the functional equivalent of a constitutional text for the EU.

12. Bundesverfassungsgericht [BVerfG] [Federal Constitutional Court] Jan. 15, 2002, 104 Entscheidungen des Bundesverfassungsgerichts [BVerfGE] 337 (F.R.G.); *see generally* Kate M. Nattrass, 2004, "'… Und Die Tiere' Constitutional Protection for Germany's Animals," 10 *Animal L.* 283; Claudia E. Haupt, 2010, "The Nature and Effects of Constitutional State Objectives: Assessing the German Basic Law's Animal Protection Clause," 16 *Animal L.* 213; Claudia E. Haupt, 2007, "Free Exercise of Religion and Animal Protection: A Comparative Perspective on Ritual Slaughter," 39 *Geo. Wash. Int'l L. Rev.* 839.

13. Interestingly, there is controversy over whether Islamic principles require killing without stunning and this was an issue raised in the decision as well. *See* L. Edwards, 2009, *Mid-East Princess in Plea to Rudd on Animal Slaughter* [online], *The Sydney Morning Herald*. Available from: http://www.smh.com.au/environment/mideast-princess-in-plea-to-rudd-on-animal-slaughter-20091101-hrkx.html [Accessed: 17.6.2010] ("The King of Jordan's sister has appealed to the Prime Minister, Kevin Rudd, to stop the ritual slaughter of conscious animals for halal meat in Australia, saying it was not necessary under Islamic principles. Princess Alia bint al-Hussein of Jordan, who is the sister of King Abdullah II of Jordan, said she had written to Mr Rudd saying that any lowering of animal welfare standards in Australia for religious reasons would be a blow to the country's reputation and undermine progress in the Middle East.").

14. Germany Gesetz zur Anderung des Tierschutzgesetzes [Animal Protection Act], May 25, 1998, BGBl. I at 1094 (F.R.G.). For an English translation of the Animal Welfare Act, see Michigan State University College of Law Animal Legal & Historical Center, Non-US Statute/Law Listing: German Animal Welfare Act, http://www.animallaw.info/nonus/statutes/stdeawa1998.htm (accessed 16.9.2010) [hereinafter German Animal Protection Act].

15. Haupt, "The Nature and Effects of Constitutional State Objectives," at 220.

16. *Ibid.*

17. Nattrass, "Constitutional Protection for Germany's Animals," at 311.

18. *Ibid.*

19. See Haupt, "The Nature and Effects of Constitutional State Objectives," at 225–226.

20. *Ibid.* at 227.

21. *Ibid.* at 230.

22. Hessischer Verwaltungsgerichtshof [State Administrative Court of Hesse], Nov. 24, 2004, 27 Natur und Recht [NuR] 464 (2005).

23. Bundesverwaltungsgericht [BVerwG] [Federal Administrative Court], Nov. 23, 2006, 26 Neue Zeitschrift für Verwaltungsrecht [NVwZ] 461 (2007).

24. See for example Haupt, "The Nature and Effects of Constitutional State Objectives," at 243.

25. See for example *Bennett* v. *Bennett*, 655 So. 2d 109 (Fla. App. 1 Dist. Jan. 19, 1995) (finding that trial court lacked authority to order dog visitation).

26. See *Raymond* v. *Lachmann*, 264 A.D. 2d 340, 341 (N.Y. Sup. Ct. 1st Div. 1999) (finding "it best for all concerned that, given his limited life expectancy, Lovey, who is now almost ten years old, remain where he has lived, prospered, loved and been loved for the past four years").

27. The Uniform Trust Act of 2000 § 408 states:

SECTION 408. TRUST FOR CARE OF ANIMAL.

(a) A trust may be created to provide for the care of an animal alive during the settlor's lifetime. The trust terminates upon the death of the animal or, if the trust was created to provide for the care of more than one animal alive during the settlor's lifetime, upon the death of the last surviving animal.

(b) A trust authorized by this section may be enforced by a person appointed in the terms of the trust or, if no person is so appointed, by a person appointed by the court. A person having an interest in the welfare of the animal may request the court to appoint a person to enforce the trust or to remove a person appointed.

(c) Property of a trust authorized by this section may be applied only to its intended use, except to the extent the court determines that the value of the trust property exceeds the amount required for the intended use. Except as otherwise provided in the terms of the trust, property not required for the intended use must be distributed to the settlor, if then living, otherwise to the settlor's successors in interest.

As of December 2009, 42 states and the District of Columbia allow for pet trusts.

28. 785 N.E.2d 811 (Oh. Ct. App. 2003).
29. *Ibid.* at 815.
30. See G. Fancione, 2005, *Animals, Property and the Law*, 2nd edn, University Press, Philadelphia, 42 (quoting *Blair* v. *Forehand*, 100 Mass. 136 (Mass. 1868) (cited in *Sentell* v. *New Orleans & C.R. Co.*, 166 U.S. 698, 703 (1897))).
31. *Petco* v. *Schuster*, 144 S.W.3d 554 (Tex. app. 2004).
32. 886 S.W.2d 368 (Tex. app. 1994)
33. 144 S.W.3d 554 (Tex. app. 2004)
34. Adoption fees for shelter animals vary. For example, in 2009 in Washington DC, adoption fees for cats at the Washington Humane Society ranged from $40 to $85.
35. 144 S.W.3d at 561.
36. *Ibid.* at 564.
37. See generally A. Karp, 2006, "Interference with Human-Companion Animal Relationship: Where 'Use' = 'Companionship' and 'Loss of Use' = 'Loss of Companionship'," *ABA-TIPS Animal Law Committee Newsletter*, Fall, 1.
38. 144 S.W.3d at 564–565.
39. Conn. Stat. § 22-351a.
40. 510 Ill. Stat. § 70/16.3, 16.5.
41. Tenn. Code § 44-17-403. Reprinted as an example.

(a)(1) If a person's pet is killed or sustains injuries that result in death caused by the unlawful and intentional, or negligent, act of another or the animal of another, the trier of fact may find the individual causing the death or the owner of the animal causing the death liable for up to five thousand dollars ($5,000) in noneconomic damages; provided, that if the death is caused by the negligent act of another, the death or fatal injury must occur on the property of the deceased pet's owner or caretaker, or while under the control and supervision of the deceased pet's owner or caretaker.

(2) If an unlawful act resulted in the death or permanent disability of a person's guide dog, then the value of the guide dog shall include, but shall

An Introduction to Animals and the Law

not necessarily be limited to, both the cost of the guide dog as well as the cost of any specialized training the guide dog received.

(b) As used in this section, 'pet' means any domesticated dog or cat normally maintained in or near the household of its owner.

(c) Limits for noneconomic damages set out in subsection (a) shall not apply to causes of action for intentional infliction of emotional distress or any other civil action other than the direct and sole loss of a pet.

(d) Noneconomic damages awarded pursuant to this section shall be limited to compensation for the loss of the reasonably expected society, companionship, love and affection of the pet.

(e) This section shall not apply to any not-for-profit entity or governmental agency, or its employees, negligently causing the death of a pet while acting on the behalf of public health or animal welfare; to any killing of a dog that has been or was killing or worrying livestock as in § 44-17-203; nor shall this section be construed to authorize any award of noneconomic damages in an action for professional negligence against a licensed veterinarian.

Chapter 6

1. See for example A. Linzey, 2009, *Why Animal Suffering Matters: Philosophy, Theology, and Practical Ethics*, Oxford University Press, New York; J. Donovan and C. Adams, 2007, *The Feminist Care Tradition in Animal Ethics*, Columbia University Press, New York; P. Singer, ed., 2006, *In Defense of Animals*, Blackwell, Oxford; R. Garner, 2005, *The Political Theory of Animal Rights*, Manchester University Press, Manchester; T. Regan, 2004, *Empty Cages: Facing the Challenge of Animal Rights*, Rowman & Littlefield, New York; C. Sunstein and M. Nussbaum, eds., 2004, *Animal Rights: Current Debates and New Directions*, Oxford University Press, New York; T. Regan, 2004, *The Case for Animal Rights*, University of California Press, 2nd edn, California; C. Cohen and T. Regan, 2001, *The Animal Rights Debate*, Rowman & Littlefield, New York; M. Rowlands, 1998, *Animal Rights: A Philosophical Defence*, Palgrave Macmillan, New York; G. Francione, 1996, *Rain Without Thunder: The Ideology of the Animal Rights Movement*, Temple University Press, Philadelphia; P. Carruthers, 1992, *The Animals Issue*, Cambridge University Press, Cambridge; P. Singer, 1990, *Animal Liberation*, 2nd edn, Random House, New York; P.A.B. Clarke and A. Linzey, eds, 1990, *Political Theory and Animal Rights*, Pluto Press, London.
2. Linzey, *Why Animal Suffering Matters*, at 133.
3. The writings of these scholars are cited such that anyone interested may explore them more fully. While each has written extensively, Sunstein and Nussbaum, *Animal Rights*, provides an excellent overview of these and others' legal views on animal rights.
4. C. Sunstein, 2004, "Introduction, What are Animal Rights?" in Sunstein and Nussbaum, *Animal Rights*, at 3–15; 2004, "Can Animals Sue?" C. Sunstein, in Sunstein and Nussbaum, *Animal Rights*, at 251–262.
5. D. Favre, 2010, "Living Property: A New Status for Animals Within the Legal System," 93 *Marq. L. Rev.* (forthcoming 2010); *see also* D. Favre, 2004, "New Property Status for Animals: Equitable Ownership," in Sunstein and Nussbaum, *Animal Rights*, at 234–250.

6. This discussion is taken from Professor Favre's most recent article, "Living Property." Favre expressly states that this theory is designed to accommodate domestic animals and leaves for another day "animals outside the realm of property law." Favre, "Living Property." Arguably, captive wild animals would fit this paradigm as well.

7. Favre includes all living beings and then, he identifies several practical limitations that will exclude some living beings and differentiate among others. Plants are excluded for lack of sufficient knowledge about them. Federal Ethics Committee on Non-Human Biotechnology (ECNH) in Switzerland, for the consideration of plants for their own sake, 2008, *The Dignity of Living Beings With Regard to Plants: Moral Consideration of Plants for Their Own Sake* [online]. Available from: http://www.scribd.com/doc/2931685/The-Dignity-of-Living-Beings-with-Regards-to-Plants [accessed: 8.21.2010]. Invertebrates are excluded in order to limit the focus on those who have the most need and for whom we can do the most. Finally, since individual identification is important, Favre distinguishes between animals with human names and others, stating: "for an animal to have a human designated name suggests a level of human concern, recognition and interaction which separates them from the unnamed living property." Favre, "Living Property."

8. Favre, "Living Property."

9. *Ibid.*

10. *Ibid.*

11. *Ibid.*

12. *Ibid.*

13. S. Wise, 2004, "Animal Rights One Step at a Time," in Sunstein and Nussbaum, *Animal Rights*, at 19–50; see also S. Wise, 2000, *Rattling the Cage: Toward Legal Rights for Animals*, Perseus Publishing, United States.

14. Wise, "Animal Rights One Step at a Time," at 19–26.

15. *Ibid.* at 32.

16. *Ibid.* at 33.

17. *Ibid.* at 40.

18. *Ibid.* at 41.

19. See generally G. Francione, *Rain Without Thunder*; G. Francione, 1997, "Animal Rights Theory and Utilitarianism: Relative Normative Guidance," 3 *Animal L.* 75; G. Francione, 2004, "Animals – Property or Persons," in Sunstein and Nussbaum, *Animal Rights*, at 108–142; G. Francione, 2008, *Animals as Persons: Essays on Abolition of Animal Exploitation,* Columbia University Press, New York.

20. See Francione, *Animals as Persons*, at 15.

21. *Ibid.* at 113–115.

22. *Ibid.* at 114.

23. *Ibid.*

24. See generally Eleanor Evertsen and Wim de Kok, 2009, "Legal Protection of Animals: The Basics," 5 *J. Animal L.* 91.

25. *Ibid.* at 97.

26. *Ibid.*

27. *Ibid.*

28. *Ibid.* at 98.

29. *Ibid.* at 97.
30. *Ibid.*
31. *Ibid.* at 101.
32. *Ibid.* at 98.
33. *Ibid.*
34. *Ibid.* at 101.
35. *Ibid.*
36. *Ibid.* at 100–101.
37. *Ibid.* at 101.
38. Linzey, *Why Animal Suffering Matters,* at 49–55.

Index

Compiled by Sarah Schott

Ringling Brothers and Barnum Bailey
  Circus, 99
Ritual slaughter. *See* Religious
  slaughter
Rivlin, Eliezer, 51, 205n112
Roberts, John G., 40
Rodeos, 26, 92–93
Roman law, 19
Royal Society for the Prevention of
  Cruelty to Animals (RSPCA), 3,
  115
Rule of law *vs.* rule of nature, 11
Rules and treaties, 12, 18–19
Rules of reason (standards), 18, 174,
  191
Russell, William, 211n36
Ryder, Richard, 3

Sacrifice, animal, 44–47
Sanctions, criminal, 15
Sankoff, Peter, 28
Santeria faith, 44–45, 47
Santorum, Rick, 213n96
Scalia, Antonin, 46
Schweitzer, Albert, 146
Seal fur, 147
Search and seizure, reasonable, 120,
  153
Secondary legislation (EU), 17
Second-generational rights, 14
Sentience. *See* Animals, sentience and
Serial killers, animal abuse and, 29
Serious injury, legal definition of,
  226n36
Sheep, used for food, 17, 107–109,
  217n174
Shelters
  adoption fees for shelter animals,
    229n34
  hold periods and, 120
  homeless companion animals in,
    122–123, 220n14
  no-kill philosophies in, 122–123,
    220n14
Singer, Peter, 3, 51
Slaughter, animal
  as art, 2
  factory farming and, 1, 36–37, 43,
    104–105, 110–111

humane slaughter laws, 46–49,
  105–110, 204n93, 204n96
Jewish traditional methods, 46–48
lack of legal protection for animals,
  108
of livestock, 108–109
methods used for, 48, 64, 107, 108,
  113, 159–160, 204n96, 204n96,
  227–228n13
religious, 46–49, 204n96,
  227–228n13
for sport, 2
stunning and, 48–113, 68, 108, 159,
  160, 227–228n13
transport, regulating, 105–108
USDA regulations for, 48, 86–87,
  99–100, 108–109, 112, 176
Slaughterhouse work, links to crime
  and, 30–31
Smith rule, 46
Snares, 60–62, 64–65, 136, 141,
  208n182
"Son of Sam" (David Berkowitz), 29
South Carolina, population
  management in, 142
Sow gestation crates, 50, 55, 110, 156
Spain
  anti-cruelty laws, 2, 193–194nn6–7,
    196n42
  bull fighting, 2
Spaying/neutering, animal, 123nn7–8
Speciesism/speciesists, 3–4, 172, 183,
  192
  non-speciesist legal system ideas, 4,
    172, 183–187, 191–192
Speech, freedom of
  animal crush videos and, 38, 41–43
  Establishment Clause, 44, 47,
    204n99
  expressive entertainment,
    202–203n66
  overview of, 14, 39–44, 140
  unprotected speech, 40, 43
Sport, animals used for. *See*
  Entertainment, animals used for;
  Hunting and trapping
Sportsmen's Caucus, 137
Standards (rules of reason), 18, 174,
  191